Hydroscience and Engineering

Series Editors

Feng Jin, Department of Hydraulic Engineering, Tsinghua University, Beijing, China

Duruo Huang, Department of Hydraulic Engineering, Tsinghua University, Beijing, China

This Springer book series Hydroscience and Engineering addresses multidisciplinary advancements in water-related science and engineering practice. Main scope of the book series includes fundamental research and cutting-edge engineering technology in water resources and river/reservoir management in relation with hydraulic structures, erosion control, sediment transport, river basin management and planning, flood control, geological/hydrological risk assessment, offshore and costal engineering. The book series focuses on developing innovative, sustainable and environment-friendly materials and structures to improve infrastructural functionality and sustainability, and to enhance adaptability to changing environmental conditions. Recent advancements and successful insights obtained from engineering practice in China and worldwide will also be thoroughly introduced.

The overarching goal of the book series is to provide combined knowledge on foremost research and solid engineering expertise on hydroscience and engineering from an interdisciplinary standpoint. The individual book volumes in the series are thematic, each specializing a focused topic, such as a new dam type and innovative dam construction materials. As a collection, the book series provides valuable resources targeting on a wide spectrum of audience, including researchers in academia, practicing engineers in engineering community and students who work on expanding their knowledge in the related areas.

Bin Xu · Rui Pang

Stochastic Dynamic Response Analysis and Performance-Based Seismic Safety Evaluation for High Concrete Faced Rockfill Dams

 Springer

Bin Xu
School of Infrastructure Engineering
Dalian University of Technology
Dalian, China

Rui Pang
School of Infrastructure Engineering
Dalian University of Technology
Dalian, China

ISSN 2730-9002   ISSN 2730-9010 (electronic)
Hydroscience and Engineering
ISBN 978-981-97-7197-4   ISBN 978-981-97-7198-1 (eBook)
https://doi.org/10.1007/978-981-97-7198-1

This work was supported by National Natural Science Foundation of China (52009017, 51979026, 52379117, 52279096).

This Springer imprint is published by the registered company Springer Nature Singapore Pte Ltd.
The registered company address is: 152 Beach Road, #21-01/04 Gateway East, Singapore 189721, Singapore

If disposing of this product, please recycle the paper.

Contents

Chapter 1
Introduction

1.1 Background

Although the western region of China is rich in hydropower resources and is suitable for the construction of large hydro projects. These projects are in the Himalayan-Mediterranean seismic belt and the geological conditions are relatively complex, the seismic intensity is high (Fig. 1.1), and the seismic activity is relatively frequent (Fig. 1.2). According to the statistics of China Earthquake Administration, more than 80% of the strong earthquakes in modern times occurred in the western region of China. Since the twentieth century, there have been nearly 70 severe earthquakes with a magnitude of more than 7 (Zhang 2017), among which the most typical are the Wenchuan earthquake in Sichuan Province in 2008 and the Yushu earthquake in 2010. The results of seismic activity and earthquake trend prediction in China show that there may be about 40 earthquakes of magnitude 7 and above and 3–4 earthquakes of magnitude 8 and above in the mainland of China in the next hundred years (Kong and Zou 2016).

Therefore, under the threat of strong earthquakes, the safety of these high dams and reservoirs must be considered as a key issue in engineering construction. So far, there are few cases of earthquake-tested concrete faced rockfill dams and related earthquake damage. Only a few concrete faced rockfill dams over 50 m in the world have been subjected to strong earthquakes. For example: the Zipingpu Dam in China (Chen et al. 2008a, b; Chen et al. 2008a, b; Guan 2009a, b; Kong et al. 2011; Kong et al. 2011; Liu et al. 2015; Ren et al. 2016; Wang et al. 2018; Yang et al. 2009; Zhang et al. 2015; Zhao et al. 2009), the Cogoti Dam in Chile (Han and Kong 1996; Noguera 1987), the Minase Dam in Japan (Han and Kong 1996), the Malpasse Dam in Peru (Han and Kong 1996), the Cogswell Dam in the United States (Boulanger et al. 1995), and the Urto Kyisk Dam (Shen 2007a, b). The main earthquake damage forms are summarized as Table 1.1 (Zhang 2017). It can be seen that the main seismic damage of concrete faced rockfill dam is as follows: the dam body has settlement and displacement to the downstream direction; local cracking, damage, void and joint

© The Author(s) 2025   1
B. Xu and R. Pang, *Stochastic Dynamic Response Analysis and Performance-Based Seismic Safety Evaluation for High Concrete Faced Rockfill Dams*, Hydroscience and Engineering, https://doi.org/10.1007/978-981-97-7198-1_1

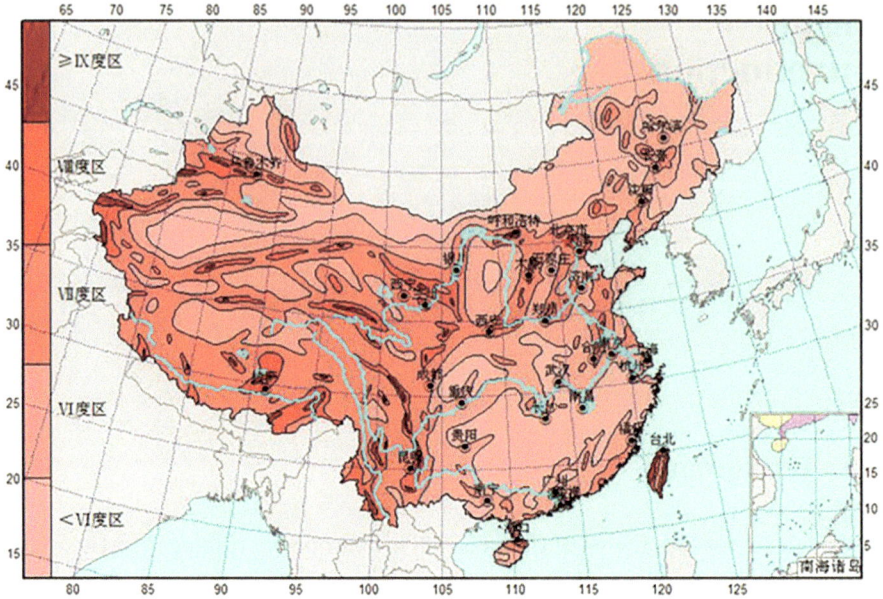

Fig. 1.1 Division map of seismic intensity in China

Fig. 1.2 Distribution of earthquakes with magnitudes greater than 4.0 in the last five years

dislocation of the panel; the downstream dam slope rock rolls down and develops into shallow slope sliding. In the Wenchuan earthquake, the typical failure mode of the Zipingpu concrete faced rockfill dam is shown in Fig. 1.3. However, it is obvious from the earthquake damage cases that most of them only exhibited light and repairable local damage without dam break.

It has shown that the dam designed and constructed in accordance with the current specification requirements has an appropriate degree of seismic capacity under earthquake action. However, as a complex natural disaster, the occurrence time, scene and intensity of earthquakes are full of randomness. The existing earthquake damage cases show that the real intensity of the dam site during the earthquake may far exceed the design intensity. For example, during the Wenchuan earthquake, the seismic intensity of the 150-m-level Zipingpu concrete faced rockfill dam was higher than its design intensity (the fortification intensity of the Zipingpu concrete faced rockfill dam project was VIII degrees. The design standard of 2% exceed probability in the base period of 100 years was adopted, and the peak acceleration of the design ground motion was 0.26 g. In the Wenchuan earthquake, the Zipingpu Water Conservancy Project is only 17.17 km away from the epicenter, and the seismic transmission intensity reaches IX–X degrees. The peak acceleration of the dam bedrock is greater than 0.5 g (Guan 2009a, b). More importantly, there is no high dam with a height of more than 200 m that has been tested by strong earthquakes, which can provide reference for seismic design and research up to now. Considering the irreplaceable important position of high dams in China, the seismic safety problems that cannot be ignored and the unpredictable secondary disasters, it is extremely important to ensure the seismic safety of high dams. Therefore, it is of great scientific significance and engineering value to study the seismic performance of high concrete faced rockfill dam under earthquake, especially under strong earthquake.

At present, the performance-based structural seismic safety design and safety evaluation methods have been widely studied in the fields of structural engineering, bridge engineering and other fields at home and abroad. However, in the field of high dams, especially high concrete faced rockfill dams, it is still in its infancy. From the analysis of the seismic performance level of the concrete faced rockfill dam, it illustrated that the randomness of the seismic load and the uncertainty of the structure itself may lead to different degrees of damage to the concrete face rockfill dam (Kartal et al. 2010; Wang et al. 2013), which will have an impact on the use function and engineering safety. Therefore, it is urgent to consider various uncertain factors under seismic action based on a variety of strong nonlinear analysis. At the same time, a reasonable evaluation method based on multi-performance indicators and different performance levels is proposed to gradually realize the transformation from deterministic analysis to uncertain probabilistic analysis. The purpose of this book is to consider the uncertainty of seismic load and structural material parameters, aiming for studying the failure probability of high concrete faced rockfill dam under earthquake by using strong nonlinear numerical analysis method. A performance-based seismic safety evaluation framework for high concrete faced rockfill dam is preliminarily established.

Table 1.1 Summary of earthquake damage situation in several CFRDs

Engineering	Year of completion	Height/m	Original time of earthquake	Earthquake magnitude	Damage condition
Cogoti Dam	1938	85	1943	8.3	Loose rocks on the crest and downstream dam slope become dislodged or even roll down. The downstream dam slope gradient changes from 1:1.5 (pre-earthquake) to 1:1.65 (post-earthquake). The dam crest experiences a settlement deformation of about 38.1 cm. Panels near the dam crest become suspended due to the settlement, leading to vertical cracks and crushing at the upper part of the panels. Opening of the surrounding joints results in vortex flow
Minase Dam	1963	66.5	1964	7.5	The dam body exhibits a horizontal displacement of 4 cm and a settlement of 6.1 cm. Cracks appear on the dam crest road surface, and there is slight damage to the panel joints with minor loss of integrity. The seepage rate increases from 90 L/s before the earthquake to 220 L/s after the earthquake
Zipingpu Dam	2006	156	2008	8.0	The maximum settlement of the dam reaches 0.81 m, which is notably significant. The downstream side of the dam experiences a horizontal displacement of over 0.3 m. Loose rock on the downstream face of the dam near the dam crest becomes dislodged and shifts. The panels show signs of compression damage, construction joint displacement, and partial detachment between the panels and the cushion layer, with a maximum offset of 17 cm. Seepage rate shows a slight increase compared to before the earthquake

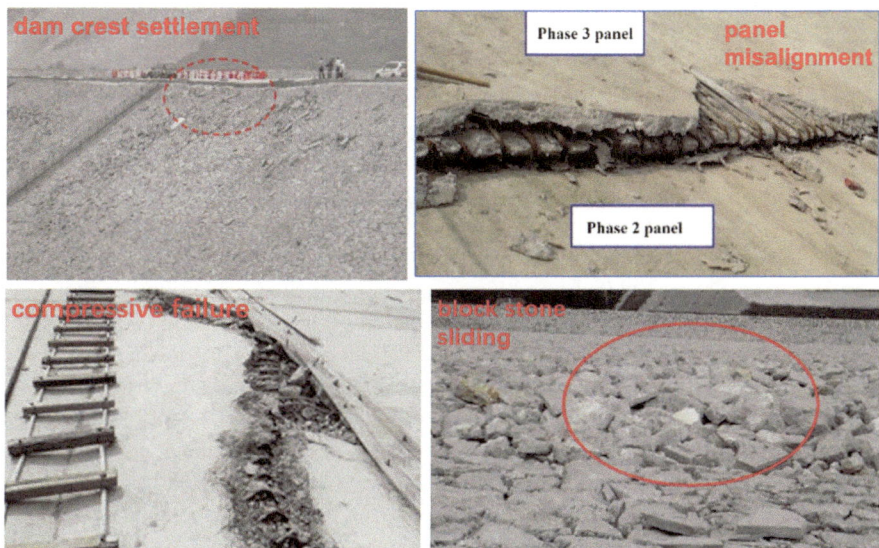

Fig. 1.3   The typical damage phenomenon of Zipingpu CFRD in Wenchuan earthquake

1.2   Performance-Based Research of Safety Evaluation of Dams

Considering the randomness and uncertainty of the seismic load and the dam structure itself, the dynamic response and failure of the dam are also random and uncertain in nature. When the earthquake occurs, the degree of structural damage caused by different probability levels is different. This will have an impact on the use function of the project and the cost of repair, and will also lead to different estimated costs and economic losses. Therefore, the analysis methods of safety assessment and failure process of high dams subjected to different intensity earthquakes need a further development. An index system which can be linked to the functional objectives of high dam seismic resistance and can express the seismic safety of high dam is constructed gradually. Finally, the unification of economy and security can be realized. According to the different materials, the dam is mainly divided into two types: concrete dam and earth-rock dam, which will show different seismic response characteristics. Zhang et al. (2016) carried out a systematic study on the theoretic of seismic design of high concrete dams. The basic framework of high dam seismic performance design based on seismic hazard analysis, dam seismic vulnerability analysis and seismic loss analysis is established.

1.2.1 Concrete Dam

Chen (2005) and Lin and Chen (2001) carried out the research work on the seismic safety evaluation, seismic fortification level and corresponding performance indexes of concrete dams. Based on the design service life and the function of the high dam, Jia and Jin (2005, 2006) proposed two methods related to the decision-making of high dam fortification standard, and pointed out that the minimum total expected loss was taken as the standard for the decision-making of optimal seismic fortification intensity. Shen (2007a, b) and Kou (2009) established a quantitative evaluation model for earthquake damage, and constructed a performance-based seismic safety and risk assessment system for high dams. Zhang et al. (2013) proposed a strategy for probabilistic analysis of the stochastic response of gravity dams based on the probabilistic analysis of dynamic response parameters of the dams. They obtained the probability distribution characteristics and patterns of gravity dams, and reasonably assessed the probabilities of various types of damage to gravity dams during earthquakes. Xu et al. (2010) constructed a model for probabilistic analysis of damage distribution in concrete gravity dams based on stochastic perturbation theory and virtual excitation method. They also developed an evaluation method for dam failure losses based on grey system dynamics. Yao et al. (2013) conducted research on seismic vulnerability analysis in the dynamic design of high arch dams and developed a performance-based seismic safety assessment method for high arch dams. Li et al. (2013) established an earthquake risk assessment approach for concrete gravity dams based on seismic hazard analysis, existing dam vulnerability analysis, and loss estimation. Li et al. (2012) utilized various artificial intelligence algorithms to construct vulnerability and risk analysis methods for concrete gravity dams. Pan et al. (2015) and Chen et al. (2019) introduced Incremental Dynamic Analysis (IDA) to perform seismic performance analysis on gravity dams and arch dams, offering a new approach for performance-based seismic design of dams, which achieved positive outcomes. Abroad, many scholars employed Incremental Dynamic Analysis and vulnerability analysis methods to analyze the seismic performance of concrete dams, exploring quantified standards for failure modes, gradually becoming a hot topic in the research of performance-based seismic safety assessment for dams, exhibiting strong novelty and practicality (Hariri et al. 2016; Hariri and Saouma 2016a; Hariri and Saouma 2016b; Hebbouche 2013; Kadkhodayan et al. 2016; Morales et al. 2016; Soysal et al. 2016; Tekie and Ellingwood 2003; Tekie 2002).

1.2.2 Rockfill Dam

Currently, both domestically and internationally, there is limited research on performance-based seismic safety assessment of earth-rock dams, particularly high-profile concrete faced rockfill dams. Apart from conducting ultimate seismic capacity analysis (Chen et al. 2013; Lu and Dou 2014; Tian et al. 2013; Wang and Zhu 2017;

Zhang and Li 2014; Zhao et al. 2015; Zhu et al. 2018), the research has mainly focused on assessing seismic safety from a probabilistic or vulnerability perspective, with some incorporation of seismic risk analysis. Wang et al. (2013, 2012) based on seismic risk analysis theory, established the evaluation indicators for relative settlement at the crest of the dam, divided the seismic damage levels of earth-rock dams, obtained risk probabilities for different damage levels through vulnerability analysis, and finally integrated seismic economic loss analysis to establish a method for seismic risk analysis of earth-rock dams. Wang et al. (2016) established vulnerability models for common failure modes (slope stability and permanent deformation) of high earth-rock dams under seismic loads, proposed a performance-based seismic risk analysis model and evaluation matrix for high earth-rock dams. Kartal et al. (2010), using the example of the Torul panel rockfill dam, employed an improved response surface method, considering uncertainties in material and geometric properties of panels and rockfill material, to analyze the crack resistance and compressive reliability of panels with different thicknesses under seismic effects. Wu et al. (2015) proposed a reliability analysis algorithm based on the generalized coordinate system, primarily used for analyzing seismic stability of dam slopes in high earth-rock dams.

From the above studies, it can be observed that for dam structures, performance-based seismic safety assessment primarily involves the analysis of various uncertain factors under seismic actions, as well as seismic response and probabilistic analysis. On the other hand, performance-based seismic safety assessment should be capable of anticipating a structure's seismic performance under potential future seismic actions. Seismic safety assessment should encompass three key points: seismic input motion, structural seismic response, and structural resistance. Therefore, "performance-based seismic safety assessment" can be summarized as ensuring that structures meet relevant seismic performance objectives under different specified seismic design levels. Its essence mainly includes: graded design standards, corresponding seismic design levels with performance objectives, and rational selection of performance indicators and quantification of performance objectives (Chen et al. 2005). Although China's current seismic design codes for hydraulic engineering (2015) partially adopt this concept, such as seismic design and verification, seismic safety assessment remains deterministic and lacks clear provisions, especially for earth-rock dams, particularly high-profile concrete faced rockfill dams. Therefore, it is necessary to conduct a seismic safety assessment of high-profile concrete faced rockfill dams from a performance probability perspective. In conclusion, the performance-based seismic safety assessment of high-profile concrete faced rockfill dams should primarily address the following issues: the true response behavior of structures under seismic actions should be reflected through effective seismic analysis models and methods; in practical applications, uncertainty factors should be thoroughly considered, and seismic response analysis should be conducted from a probabilistic standpoint; quantified performance objectives and rational performance indicators are prerequisites and foundations for seismic performance assessment. Consequently, the following section mainly provides a concise overview of the research trends in concrete faced rockfill dam studies based on these three key concerns of designers.

1.2.3 Performance-Based Seismic Design Framework

The performance-based seismic design framework for structural engineering can be succinctly summarized as follows: considering uncertainty factors under seismic actions, the structural design is tailored to meet various functional requirements, achieving an optimal balance between safety and economy. Furthermore, for complex engineering structures, the performance-based seismic safety assessment primarily encompasses three aspects: (1) analysis of various uncertain factors under seismic actions; (2) seismic response and probabilistic analysis; (3) seismic loss analysis. This process can be illustrated using Fig. 1.4 as a schematic (taking a high earth-rock dam as an example). Uncertainty factors encompass seismic motion and other load uncertainties, material parameter uncertainties, model uncertainties, and other uncertainties, etc. Seismic response and probabilistic analysis mainly involve conducting a series of response analyses for the structure under a range of seismic excitations, considering the aforementioned uncertainties, selecting appropriate performance indicators, and subsequently obtaining the probabilities of achieving different performance objectives. Seismic loss analysis refers to the calculation of various direct and indirect economic losses (including casualties) under different performance objective probabilities. However, the theory of performance-based seismic design is still in its early stages within the field of hydraulic engineering construction. Due to the significant complexity of dams, currently, no country in the world has utilized performance design theory for seismic design of dams. Therefore, it is necessary to conduct performance-based seismic safety assessments based on current conditions. Scholars in certain countries and regions have initiated relevant efforts or transitioned towards performance-based design (Chen 2010, 2005; Lin 2005, 2004; Zhang 2016).

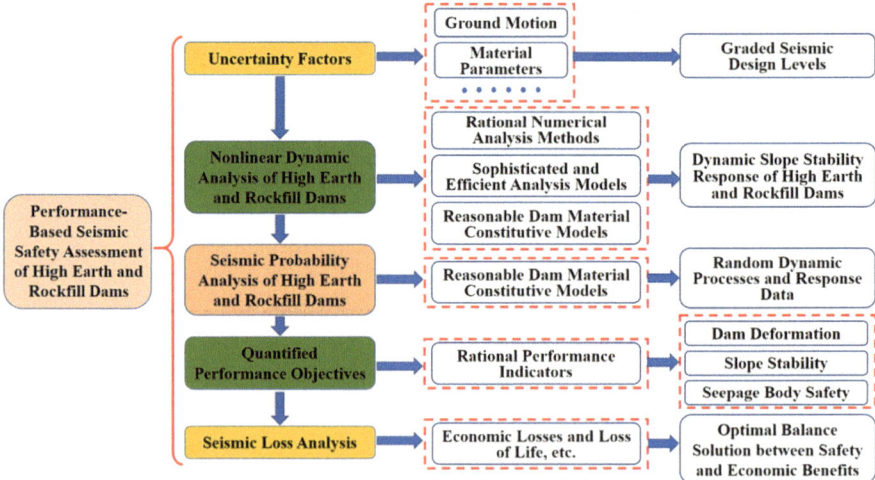

Fig. 1.4 Flow chart of performance-based seismic safety evaluation

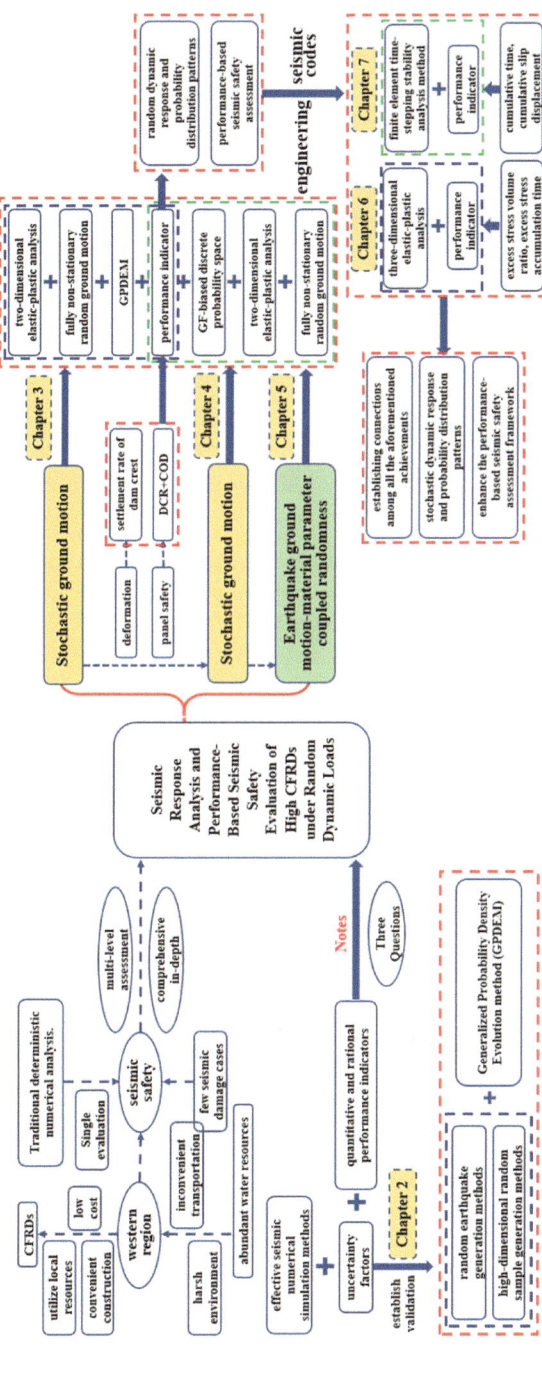

Fig. 1.5 The research framework

1.3   The Focus and Content of This Book

As mentioned earlier, the theory of performance-based seismic design has gradually been applied and developed in various engineering fields. However, for earth-rock dams, the current seismic safety assessment mainly relies on traditional deterministic analysis methods for simulation. Despite the initial forays into performance-based seismic safety assessment, especially concerning high concrete faced rockfill dams, research in this area remains relatively scarce. Current studies have also inadequately accounted for uncertainty factors under seismic conditions and have not conducted seismic safety analysis from a probabilistic perspective. Additionally, there is a lack of a systematic evaluation framework and system. Therefore, this section summarizes the existing research efforts, providing an overview of the main issues currently present. Addressing these issues, the main research content of this paper is introduced.

1.3.1   The Current Existing Problems

Based on three aspects of performance-based structural seismic safety assessment and considering the challenges posed by current disaster losses and impacts (Chen et al. 2010), this paper primarily addresses the uncertainties in seismic motion and material parameters under earthquake conditions, conducting seismic response and probabilistic analysis. The main issues are as follows:

(1) Study on describing uncertainty of seismic motion and dam-building material parameters, as well as sample selection and generation. In seismic response analysis of high concrete faced rockfill dams, uncertainties exist in both seismic motion and dam-building materials. Currently, there is limited research that adequately considers the uncertainties in seismic motion and dam-building material parameters, especially the randomness of material parameters. Furthermore, the coupled stochastic effects of these two factors have not been effectively considered. Due to the high-dimensionality and diverse statistical distributions of stochastic parameters, commonly used methods for generating stochastic samples are not suitable for large and highly nonlinear high concrete faced rockfill dams. Therefore, it is necessary to comprehensively consider the influences of seismic motion randomness, dam-building material parameter uncertainty, and the coupled randomness of seismic motion and material parameters. This involves establishing a stochastic seismic motion model and a method for generating high-dimensional stochastic parameters. These steps are essential for conducting random dynamic and probabilistic analysis of high concrete faced rockfill dams.

(2) Research on stochastic dynamic response and probabilistic analysis methods. Currently, there is limited research related to probabilistic analysis of earth-rock dams, especially concerning stochastic dynamic time history analysis. Traditional probabilistic analysis methods such as the first-order second-moment

method, Monte Carlo method, response surface method, and their improved forms may struggle to acquire the stochastic dynamic information of structures. They might involve extensive computations, coupling with structural response analysis, continuous sample training, and iteration. These challenges make them less suitable for seismic stochastic dynamic response and probabilistic analysis of highly nonlinear, complex, and computationally extensive high concrete faced rockfill dams. As a result, it's essential to develop rational and efficient probabilistic analysis methods.

(3) Research on refined stochastic dynamic time history analysis method based on elastic–plastic behavior and inconsistent seismic input. Currently, the stochastic dynamic and probabilistic analysis of earth-rock dams is generally based on equivalent linear and consistent seismic input, or quasi-static methods. However, under seismic actions, especially strong earthquakes, high concrete faced rockfill dams exhibit strong nonlinear characteristics, and research indicates that inconsistent seismic input significantly affects their dynamic response. Therefore, it's necessary to consider elastic–plastic behavior and inconsistent seismic input to conduct refined stochastic dynamic and probabilistic analysis of high concrete faced rockfill dams.

(4) Selection of performance indicators and quantification of performance objectives, as well as research on performance-based seismic safety evaluation of high concrete faced rockfill dams. Currently, a consensus has been reached regarding the comprehensive evaluation of the seismic safety of high concrete faced rockfill dams from three aspects: dam deformation, dam slope stability, and seepage prevention. However, reasonable performance indicators and quantified performance objectives have not been systematically studied and discussed. Furthermore, based on these three aspects, there is limited research from the perspectives of stochastic dynamics and probability. Nevertheless, this is crucial for the performance-based seismic safety assessment of high concrete faced rockfill dams. Therefore, introducing the concept of performance-based seismic design, after thorough discussions on the selection of performance indicators and quantification of performance objectives, a framework for the performance-based seismic safety evaluation of high concrete faced rockfill dams is formulated.

1.3.2 Main Ideas and Tasks

In response to the aforementioned main issues, a performance-based seismic safety evaluation framework for high concrete faced rockfill dams is established, systematically considering the uncertainties under seismic actions. Addressing seismic motion randomness, dam-building material parameter uncertainty, and the coupled randomness of seismic motion and material parameters, a methodology is developed based on the hydraulic seismic design spectrum for generating stochastic seismic motions. High-dimensional stochastic parameter sample generation methods and seismic motion-material parameter coupled stochastic sample generation methods are established. Combining refined nonlinear finite element dynamic time history analysis methods, probability density evolution methods, and vulnerability analysis methods, the stochastic dynamic response characteristics of high concrete faced rockfill dams are studied from a probabilistic perspective.

Proposed seismic safety evaluation performance indicators for high concrete faced rockfill dams are suggested, along with corresponding performance levels that have probabilistic guarantees. Ultimately, a multi-seismic intensity-multi-performance objective-failure probability performance relationship is established. This forms a preliminary performance-based seismic safety evaluation framework that provides a scientific basis for the seismic design and performance control of high concrete faced rockfill dams. The main contents of this book encompass the following aspects:

This Chapter: This chapter provides a brief overview of the research background and significance of this paper. It introduces the content and development of performance-based structural seismic safety design and elaborates on the key issues in the seismic safety evaluation of high concrete faced rockfill dams based on performance criteria. It points out the main challenges existing in current research and introduces the research scope of this paper.

Chapter 2: This chapter briefly outlines the uncertain factors present in earth-rock dams under seismic actions and the main probabilistic analysis methods. It focuses on introducing the generalized probability density evolution method and its relevant application process. The process of generating stochastic seismic motion and high-dimensional stochastic samples is established. The effectiveness and reliability of the generalized probability density evolution method applied to large-scale geotechnical engineering are verified. This lays the theoretical foundation for subsequent analyses of stochastic seismic response and performance-based seismic safety evaluation of high concrete faced rockfill dams.

Chapter 3: This chapter comprehensively considers the randomness of seismic excitation, combining the generalized probability density evolution method with elastoplastic analysis. From the perspectives of stochastic dynamics and probability, it reveals the seismic response patterns of high concrete faced rockfill dams, forming the foundation for a performance-based seismic safety evaluation. The stochastic dynamic and probabilistic responses of several commonly used response variables in high concrete faced rockfill dams, including dam body acceleration, deformation, and panel stress, are examined. The numerical distribution ranges of these response indicators are studied from the viewpoints of stochastic dynamics and probability under various seismic intensities. Finally, based on performance indicators that combine dam top settlement deformation and the ratio of panel demand stress, along with cumulative over-stress duration, a preliminary performance-based seismic safety evaluation framework tailored to high concrete faced rockfill dams is established.

Chapter 4: This chapter combines the high-dimensional stochastic parameter sampling method based on the GF-deviation resampling technique with the generalized probability density evolution method. It uncovers the stochastic dynamic response and probabilistic characteristics of high concrete faced rockfill dams influenced by random factors in material parameters. Using the GF-deviation resampling technique for optimized point selection, elastoplastic stochastic parameter samples are generated. These samples are then combined with elastoplastic analysis for high concrete faced rockfill dams. The chapter delves into the stochastic dynamic response and probabilistic characteristics of high concrete faced rockfill dams under deterministic seismic actions, considering the impact of random factors in material parameters. The effects of different distribution types of stochastic parameters are also compared.

Chapter 5: This chapter systematically considers the coupled randomness of seismic motion and material parameters. It thoroughly investigates the impact of this coupling on the dynamic response and seismic safety of high concrete faced rockfill dams from the perspectives of stochastic dynamics and probability. This chapter further refines the performance-based seismic safety evaluation framework. By combining spectral representation-random function methods with random material parameter variables, both stochastic seismic motions and random material parameter samples are simultaneously generated. From the viewpoints of stochastic dynamics and probability, the chapter contrasts the effects of various stochastic factors, including seismic motion randomness, material parameter uncertainty, and the coupling of seismic motion and material parameters, on the seismic response of high concrete faced rockfill dams. The framework is extended to encompass a multi-seismic intensity-multi-performance objective-exceed probability performance relationship and fragility curves considering the coupled randomness of seismic motion and material parameters under different seismic intensity levels. This finalizes the refinement of the performance-based seismic safety evaluation framework.

Chapter 6: This chapter delves into the stochastic dynamic response patterns of three-dimensional high concrete faced rockfill dams. It primarily explores the selection of performance indicators and performance levels for panel safety assessment. It establishes a connection with the aforementioned performance safety evaluation framework and further enhances the performance-based seismic safety evaluation framework. This chapter serves as a scientific basis for the seismic design and performance control of high concrete faced rockfill dams.

Chapter 7: Departing from the perspective of current hydraulic seismic codes and engineering applications, this chapter systematically considers the randomness of seismic motion, material parameter uncertainty, and the coupling of seismic motion and material parameters. It explores a performance-based seismic safety evaluation framework for the slope stability of high concrete faced rockfill dams. Using the dynamic finite element time history method for slope stability analysis, coupled with soil softening strength change calculations, and incorporating the generalized probability density evolution method, the chapter evaluates the influence of rockfill material softening characteristics on the seismic safety stability of dam slopes from stochastic and probabilistic viewpoints. By considering safety factors, the cumulative time of safety factor exceeding limits, and cumulative slip displacement, the impact of material softening on dam slope seismic safety stability is assessed. Finally, a progressive development of probabilistic analysis methods for dam slope stability and a performance-based seismic safety evaluation framework for dam slope stability are established.

Chapter 8: Conclusion and Future Outlook. This chapter summarizes the research conducted in this paper, elucidates the main points of innovation, and outlines the primary directions and content for future research endeavors (Fig. 1.5).

The finite element static, dynamic, and stability calculations utilize the independently developed software tools, GEODYNA and FEMSTABLE 2.0, by the Institute of Engineering Seismic Research, School of Hydraulic Engineering, Dalian University of Technology. This software suite, supported by funding from more than 10 projects by the National Natural Science Foundation of China, incorporates numerous advantages and advanced constitutive relationships from various foreign geotechnical engineering analysis programs like FEMDAM, QUAD8, GEOSLOPE, and FLUSH. It encompasses over ten types of elements, including continuous block elements, interface elements, beam elements, column elements, mass elements, and boundary elements (viscous boundaries). This suite is developed using the Visual C++ platform, object-oriented design methods, and advanced technologies such as CPU + GPU parallel computing. Presently, this software suite has been applied in seismic calculations and analysis for dozens of major earth-rock dam projects, nuclear power plant projects, as well as significant water transportation projects such as ports, both domestically and internationally. It has gained extensive usage and accumulated rich engineering experience.

References

Arrau L, Ibarra I, Noguera G (1987) Performance of Cogoti dam under seismic loading. J Geotech Eng 113:1136–1138

Boulanger RW, Bray JD, Merry SM, Mejia LH (1995) Three-dimensional dynamic response analyses of Cogswell Dam. Can Geotech J 32(3):452–464

Chen DH, Yang ZH, Wang M et al (2019) Seismic performance and failure modes of the Jin'anqiao concrete gravity dam based on incremental dynamic analysis. Eng Fail Anal 100:227–244

Chen HQ (2005) Seismic fortification levels and performance objectives for large dams. In: Earthquake resistance engineering and retrofitting, pp 7–12

Chen HQ (2010) Study on seismic fortification standard of hydraulic structures. China Water Resour 4–6+3

Chen HQ, Xu ZP, Li M (2008) Wenchuan Earthquake and seismic safety of large dams. J Hydraul Eng 1158–1167

Chen SS, Huo JP, Zhang WM (2008) Analysis of effects of "5.12" Wenchuan earthquake on Zipingpu concrete face rock-fill dam. Chin J Geotech Eng 795–801

Chen SS, Li GY, Fu ZZ (2013) Safety criteria and limit resistance capacity of high earth-rock dams subjected to earthquakes. Chin J Geotech Eng 35(Suppl):59–65

China Electric Power Press (2015) Code for seismic design of hydraulic structures of hydropower project, NB 35047. China Electric Power Press, Beijing

Guan ZC (2009a) Investigation of the 5.12 Wenchuan earthquake damages to the Zipingpu water control project and an assessment of its safety state. Sci China Technol Sci 52:820–834

Guan ZC (2009b) Seismic damage investigation and safety status review of Zipingpu water conservancy project 5.12. Science China Technol Sci 39:1291–1303

Han GC, Kong XJ (1996) Aseismatic studies of concrete faced rockfill dams: state-of-the-art. J Dalian Univ Technol, pp 74–86

Hariri-Ardebili MA, Saouma VE (2016a) Probabilistic seismic demand model and optimal intensity measure for concrete dams. Struct Saf 59:67–85

Hariri-Ardebili MA, Saouma VE (2016b) Sensitivity and uncertainty quantification of the cohesive crack model. Eng Fract Mech 155:18–35

Hariri-Ardebili MA, Saouma VE (2016c) Collapse fragility curves for concrete dams: comprehensive study. J Struct Eng 142:04016075

Hariri-Ardebili MA, Saouma VE, Porter KA (2016) Quantification of seismic potential failure modes in concrete dams. Earthquake Eng Struct Dynam 45:979–997

Hebbouche A, Bensaibi M, Mroueh H (2013) Seismic fragility and uncertainty analysis of concrete gravity dams under near-fault ground motions. Int Conf Civil Eng Transp

Jia C, Jin F, Wang PJ, Zhang C (2005) Design earthquake level analysis for high dams. Earthq Eng Eng Vib 4:155–158

Jin F, Jia C, Wang PJ, Zhang C (2006) Study on performance-based risk decision-making for high dam construction cases. Rock Soil Mech 27:1421–1424

Kadkhodayan V, Aghajanzadeh SM, Mirzabozorg H (2016) Seismic assessment of arch dams using fragility curves. Civil Eng J 1

Kartal ME, Bayraktar A, Başağa HB (2010) Seismic failure probability of concrete slab on CFR dams with welded and friction contacts by response surface method. Soil Dyn Earthq Eng 30:1383–1399

Kong XJ, Zou DG (2016) Earthquake disaster simulation and engineering application of high earth-rock dam. Science Press, Beijing

Kong XJ, Zhou Y, Zou DG et al (2011) Numerical analysis of dislocations of the face slabs of the Zipingpu concrete faced rockfill dam during the Wenchuan earthquake. Earthq Eng Eng Vib 10:581–589

Kou LH (2009) Research on key issues of performance-based seismic design of high dams (Doctoral dissertation). Tsinghua University

Li HJ (2012) Study on calculating methods of dam risk analysis (Doctoral dissertation). Dalian University of Technology

Li M (2013) Performance-based seismic fragility analysis and risk evaluation of concrete gravity dam (Master's thesis). Tianjin University

Lin G (2004) Development status and prospect of seismic technology of concrete dam I. Water Sci Eng Technol 1–3

Lin G (2005) Development status and prospect of seismic technology of concrete dam II. Water Sci Eng Technol 1–3

Lin G, Chen JY (2001) Seismic safety evaluation of large concrete dams. J Hydraul Eng 8–15

Liu HL, Chen YM, Yu T et al (2015) Seismic analysis of the Zipingpu concrete-faced rockfill dam response to the 2008 Wenchuan, China, earthquake. J Perform Constr Facil 29:04014129

Lu YP, Dou XX (2014) Analysis on ultimate seismic capacity and safety of ultra-high CFRD. Yangtze River 45:46–50

Morales-Torres A, Escuder-Bueno I, Altarejos-García L et al (2016) Building fragility curves of sliding failure of concrete gravity dams integrating natural and epistemic uncertainties. Eng Struct 125:227–235

Pan JW, Xu YJ, Jin F (2015) Seismic performance assessment of arch dams using incremental nonlinear dynamic analysis. Eur J Environ Civ Eng 19:305–326

Ren AW, Wang YJ, Chen ZY et al (2016) Performance of the reinforced right abutment slope of Zipingpu Dam during magnitude 8.0 earthquake, Wenchuan, China. Q J Eng GeolHydrogeol 49:298–307

Shen CJ (2007) Dynamic testing and antiseismic analysis for the Xiatianji rock-fill impervious face dam (Master's thesis). Xian University of Technology

Shen HZ (2007) Performance-based seismic damage analysis and risk evaluation model for concrete dam-foundation system (Doctoral dissertation). Tsinghua University

Soysal BF, Binici B, Arici Y (2016) Investigation of the relationship of seismic intensity measures and the accumulation of damage on concrete gravity dams using incremental dynamic analysis: damage Accumulation on Concrete Gravity Dams. Earthq Eng Struct Dyn 45(5):719–737

Tekie PB (2002) Fragility analysis of concrete gravity dams. The Johns Hopkins University

Tekie PB, Ellingwood BR (2003) Seismic fragility assessment of concrete gravity dams. Earthq Eng Struct Dyn 32(15):2221–2240

Tian JY, Liu HL, Wu XY (2013) Evaluation perspectives and criteria of maximum aseismic capability for high earth-rock dam. J Disas Prevent Mitigat Eng 33:128–131+137

Wang J, Zhu S (2017) Analysis on ultimate seismic capacity of high Lawa concrete face rockfill dam. Water Resour Hydropower Eng 48:28–34

Wang DB, Liu HL, Yu T (2012) Seismic risk analysis of earth-rock dam based on deformation. Rock Soil Mechan 33:1479–1484

Wang DB, Liu HL, Yu T et al (2013) Seismic fragility analysis for earth-rock dams based on deformation. Chin J Geotech Eng 35:814–819

Wang Q, Zhu S, Feng YM (2016) Seismic risk analysis of high earth-rock dam based on performance. Water Power 42:57–60

Wang Q, Zhu S, Feng YM (2018) Seismic risk analysis of Zipingpu concrete face rockfill dam. Water Power 44:48–51

Wu ZY, Chen JK, Li YL et al (2015) An algorithm in generalized coordinate system and its application to reliability analysis of seismic slope stability of high rockfill dams. Eng Geol 188:88–96

Xu Q (2010) Research on dynamical system's reliability for concrete gravity dam (Doctoral dissertation). Dalian University of Technology

Yao XW (2013) Performance-based seismic fragility analysis and safety assessment of high arch dams (Doctoral dissertation). Zhejiang University

Yang ZY, Zhang JM, Gao XZ (2009) A primary analysis of seismic behavior and damage for Zipingpu CFRD during Wenchuan earthquake. Water Power 35:30–33+59

Zhang BY, Li DY (2014). Review on study of ultimate aseismic capacity of high dams. Water Resour Power 32:63–65+189

Zhang CH, Jin F, Wang JT et al (2016) Key issues and developments on seismic safety evaluation of high concrete dams. J Hydraul 47:253–264

Zhang Y (2017) Research on seismic response, damage mechanism and anti-seismic countermeasure of face slab of high concrete face rockfill dam (Doctoral dissertation). Dalian University of Technology

Zhang SR, Wang C, Sun B (2013) Probabilistic characteristics of the performance-based seismic response of concrete gravity dams. J Tianjin Univ Technol 46:603–610

Zhang JM, Yang ZY, Gao XZ et al (2015) Geotechnical aspects and seismic damage of the 156-m-high Zipingpu concrete-faced rockfill dam following the Ms 8.0 Wenchuan earthquake. Soil Dyn Earthq Eng 76:145–156

Zhao JM, Liu XS, Wen YF et al (2009) Analysis of earthquake damage of the Zipingpu dam in Wenchuan earthquake and the study proposal on the anti-earthquake and disaster reduction of high earth-rock dam. Water Power 35:11–14

Zhao JM, Liu XS, Yang YS et al (2015) Criteria for seismic safety evaluation and maximum aseismic capability of high concrete face rockfill dams. Chin J Geotech Eng 37:2254–2261

Zhu KB, Liu HJ, Liu XS et al (2018) Discussion of the ultimate seismic capacity of high concrete face rock-filled dam. J Nat Disas 27:142–147

Chapter 2
Probability Analysis Method of Seismic Response for Earth-Rockfill Dams

Seismic response probabilistic analysis for earth-rock dams is a crucial step in performance-based seismic safety evaluation. It represents a significant transition from deterministic analysis to stochastic analysis. To conduct such an analysis, it's essential to thoroughly consider the uncertainties associated with the seismic response of earth-rock dams and select appropriate and effective probabilistic analysis methods. In the following, the uncertainties present in the seismic response of earth-rock dams and the primary probabilistic analysis methods will be briefly outlined. Additionally, the theoretical foundation and solution procedures of these methods will be briefly introduced. This will establish the theoretical groundwork for subsequent analyses of stochastic dynamic responses and performance-based seismic safety evaluations for high concrete faced rockfill dams.

2.1 Uncertainties in the Seismic Response of Earth-Rock Dams

In geotechnical engineering, uncertainties primarily include objective uncertainty and subjective uncertainty. Objective uncertainty is essentially determined by factors such as geotechnical engineering loads, soil parameters, various construction environments, and conditions. On the other hand, subjective uncertainty arises due to the limited human understanding of geotechnical engineering analysis and simulation. This includes aspects like computational models, assumption conditions, and simplifications. Considering the specific context of high concrete faced rockfill dams, two main uncertainty factors are primarily taken into account: the stochastic nature of seismic motion and the uncertainty in the parameters of the rockfill materials.

© The Author(s) 2025
B. Xu and R. Pang, *Stochastic Dynamic Response Analysis and Performance-Based Seismic Safety Evaluation for High Concrete Faced Rockfill Dams*,
Hydroscience and Engineering, https://doi.org/10.1007/978-981-97-7198-1_2

2.1.1 Randomness of Ground Motion

A considerable number of actual seismic records have indicated the significant uncertainty in seismic motion characteristics. Therefore, a very limited number of computational samples can hardly capture the influence of various factors related to seismic motion stochastic characteristics on the response of earth-rock dam structures, such as the frequency content, amplitude variations, peak values, duration, and arrangement order of different-amplitude pulses within the seismic motion. As early as the 1990s, scholars began exploring the stochastic dynamic response and reliability of earth-rock dams under stochastic seismic excitation. For example, Yu et al. (1993) generated 100 artificial earthquake waves through random simulation, introduced the concept of a state point, employed dynamic programming to locate the most probable sliding surface, and calculated the probability of permanent displacement of earth-rock dams. Chen et al. (1995) used a power function to describe the variation of average shear modulus with dam height and proposed a simplified method for analyzing the stochastic seismic response of heterogeneous earth dams based on a one-dimensional shear beam model. Liu (1996) and Liu et al. (1996) simulated the seismic process as a stationary Gaussian filtered white noise process, utilized the equivalent nodal force method, conducted random seismic response analysis, established an analysis method for the average permanent deformation failure probability of earth-rock dams, and developed a nonlinear random response and dynamic reliability analysis method based on the theory of random vibration and the virtual excitation method. They validated the method's rationality through numerical examples related to earth-rock dams. Shao et al. (1999) simulated the seismic action process as a zero-mean stationary Gaussian process, conducted random seismic response analysis to calculate the stochastic dynamic response of earth-rock dams, and employed the Hook-Jeeves search method to locate the most dangerous sliding plane and the minimum safety factor in the mean sense. This was compared with shake table model tests to validate the effectiveness and rationality of the approach. Wang et al. (2006) proposed a simple seismic motion model that simulates stationary random processes based on stochastic process theory. They investigated the seismic response characteristics of an actual homogeneous earth dam under random load excitation.

2.1.2 Uncertainty of Rockfill Material Parameters

Earth-rock dams are constructed using natural materials, and their properties vary naturally, making them complex with diverse physical and mechanical characteristics. Additionally, their gradation ranges widely, and on-site construction is primarily controlled by the void ratio. It's challenging to precisely control gradation, leading to significant variability and uncertainty in deformation and strength parameters of the dam materials (Wichtmann and Triantafyllidis 2013). However, both deformation and strength parameters have a substantial impact on the numerical analysis results of seismic dynamic responses for high-faced rockfill dams. Currently, in the seismic probabilistic analysis of earth-rock dams, most efforts have been directed towards the uncertainty of seismic motion, with fewer studies investigating the influence of uncertainty in dam construction materials on dam dynamic responses. Zhang and Liu (1994) collected physical and mechanical test data for construction materials from 95 dam engineering projects in China. They mainly established a statistical database based on c and φ parameters and developed a comprehensive probability statistical analysis program, resulting in many valuable outcomes. Wu et al. (1991) considered the uncertainty of rockfill mass, dynamic modulus, and damping. They combined frequency domain analysis and perturbation method in structural dynamic response analysis to study the variation of dam slope displacement. Sanchez et al. (2014) utilized the Karhunen–Loeve expansion method to simulate the stochastic fields of rockfill and dam foundation materials, and solved for the random slope displacement of an earth-rock dam. Kartal et al. (2010) using the example of the Torul rockfill dam, applied an improved response surface method to consider the uncertainty of material and geometric properties of panels and rockfill. They explored the crack resistance and compression reliability of panels with different thicknesses under seismic actions. Wang et al. (2013) combined the mechanical parameter samples of dam construction materials using orthogonal design. Using dam crest settlement as an evaluation indicator, they studied the vulnerability of earth-rock dams. Yang and Zhu (2016) proposed a combined technique of stochastic field simulation and finite element method in the stochastic finite element method. They discussed the effects of spatial uncertainty in dry density, void ratio, coefficient of non-uniformity, average particle size, Duncan-Chang E-B model modulus coefficient, and initial friction angle on the random seismic response and permanent deformation of earth-rock dams.

2.2 Probabilistic Analysis Method

Probability analysis is an effective approach for addressing uncertainty in seismic engineering of earth-rock dams and forms the foundation for the performance-based seismic safety assessment of high concrete faced rockfill dams. Traditional engineering structural probability analysis methods mainly include the first-order second-moment method, the Monte Carlo method, and the response surface method, while other methods are mostly improvements or developments based on these categories. Additionally, for probability analysis based on random vibrations, the Generalized Probability Density Evolution (GPDE) method is a recently developed innovative approach. Below, we will elucidate their concepts and developments from aspects including failure probability definition, first-order second-moment method, Monte Carlo method, and response surface method. Furthermore, we will emphasize the Generalized Probability Density Evolution method, which serves as a probability analysis tool in this paper.

2.2.1 Failure Probability Definition

Generally speaking, the factors influencing the structural performance requirements can be represented by two random comprehensive variables, namely the comprehensive effect S of the structure and the comprehensive resistance R of the structure. Therefore, the functional expression of a structure reaching a certain limit resistance can be quantified as:

$$Z = R - S \tag{2.1}$$

Therefore, Z is also a random variable. When $Z > 0$, the structural response is within the safe region, indicating that the functional safety requirements are satisfied, and this quantified indicator is represented by P_f. When $Z < 0$, the structural response is outside the safe region, signifying structural functional failure, and this quantified indicator is represented by P_f. Since the factors influencing S and R are a series of more fundamental random variables (such as the strength parameters of earth-rock dams, seismic loads, displacement responses, etc.), let these fundamental random variables be $X_1, X_2, \ldots, X_n$. Then, the general form of the functional expression can be represented as:

$$Z = g(X_1, X_2, \ldots, X_n) \tag{2.2}$$

The limit state equation is:

$$Z = g(X_1, X_2, \ldots, X_n) = 0 \tag{2.3}$$

2.2.2 First-Order Second-Moment Method

The basic principle of the First-order Second-moment Method (FOSM) (Hasofer and Lind 1974) is to expand the limit state function at a specific point using a Taylor series, considering the uncertainty of random variable distributions. The linearized first-order term is selected, and the mean and standard deviation of the random variables are used to calculate reliability indices. Expanding the limit state function at the point $X_{0i} (i = 1, 2, \ldots, n)$ using a Taylor series, we have:

$$Z = g(X_{01}, X_{02}, \ldots, X_{0n}) + \sum_{i=1}^{n} (X_i - X_{0i}) \left(\frac{\partial g}{\partial X_i} \right)_{X_0}$$
$$+ \sum_{i=1}^{n} \frac{(X_i - X_{0i})^2}{2} \left(\frac{\partial^2 g}{\partial X_i^2} \right)_{X_0} + \cdots \tag{2.4}$$

In order to derive the linear limit state equation, only the first-order term is approximated:

$$Z \approx g(X_{01}, X_{02}, ..., X_{0n}) + \sum_{i=1}^{n} (X_i - X_{0i}) \left(\frac{\partial g}{\partial X_i} \right)_{X_0} \tag{2.5}$$

$\left(\frac{\partial g}{\partial X_i} \right)_{X_0}$ represents the value of that derivative at point $X_{0i} (i = 1, 2, ..., n)$. Equation (2.5) is a commonly used formula for linearizing the functional expression in reliability analysis.

From the above fundamental principles, it can be seen that the first-order second-moment method, including its various improved forms, is straightforward and computationally convenient. However, for complex nonlinear problems, its accuracy is difficult to ensure. Convergence of iterations is often problematic, and it is no longer applicable to engineering structures that frequently involve implicit functions or complex state functions. In conclusion, it is challenging to apply this method to probabilistic analysis of large and complex nonlinear structures.

2.2.3 Monte Carlo Method

The Monte Carlo method is a numerical computation approach that involves generating random samples of variables based on different probability distribution characteristics. These random variable samples are then used as inputs to obtain samples of a functional function. By counting the number of samples that result in failure or destruction, the method estimates the probability of failure. This approach has a clear conceptual framework, is easy to use, and finds wide application in the field of probability analysis. As the number of simulations increases, the accuracy of the Monte Carlo method improves, but the computational complexity also increases significantly. This makes it challenging to apply to practical engineering scenarios, especially for large-scale and strongly nonlinear projects such as high earth-rock dams. Nonetheless, it is often used to validate the accuracy of other new probability analysis methods.

To enhance the efficiency of numerical simulation methods, various techniques have been developed based on the Monte Carlo foundation, such as importance sampling, subset simulation, line sampling, directional sampling, and Latin hypercube sampling. While these methods improve computational efficiency, they require the determination of design points for limit state functions to obtain important density functions. However, obtaining design points for limit states can be difficult in practical engineering, especially for complex projects, presenting significant challenges to widespread adoption. The estimated probability of failure obtained using the Monte Carlo method can be summarized briefly as follows:

$$P_{\mathrm{f}} = \frac{n_{\mathrm{f}}}{n} \tag{2.6}$$

in which, n_{f} represents the number of samples in the failure region, and n is the total number of samples.

2.2.4 Response Surface Methodology

For large and complex structures, the limit state function is often implicit. The response surface methodology (RSM) involves constructing a simple explicit function that progressively approximates the actual implicit (or explicit) limit state function. This simplification aids in the calculation of probabilistic reliability and has gained increasing attention in recent years. In 1951, Box and Wilson (1951) first introduced the response surface methodology, focusing on how to use statistical methods to obtain an explicit function for approximating a complex implicit function. Wong (1985) was the first to apply the response surface methodology to analyze the reliability of slope stability. Subsequently, numerous scholars both domestically and internationally have further expanded the research related to this methodology. Based on random variables, the form is as follows:

$$Y' = g'(X) = a + X^T B + X^T C X \tag{2.7}$$

Determining the constant a and constant matrices B and C requires an adequate number of sample points.

Regarding the reliability analysis of structures, the associated random variables are often numerous. To determine the unknown coefficients in Eq. (2.7), a substantial number of sample points need to be analyzed and computed. This can impact the computational efficiency of the response surface methodology. In order to achieve both sufficient computational accuracy and improved efficiency, in 1990, Bucher and Bourgund (1990) introduced the quadratic response surface function, which has been widely utilized. Its form is as follows:

$$g'(X) = a + \sum_{i=1}^{n} b_i X_i + \sum_{i=1}^{n} c_i X_i^2 \tag{2.8}$$

In the equation: coefficients a, b_i, and c_i need to be determined by obtaining a sufficient number of equations using $2n + 1$ sample points. However, since the interaction terms do not appear in the response surface function, the number of sample points required for determining the response surface function decreases. The specific calculation method is as follows:

The first step involves considering the mean as the central point and selecting sample points within the interval $(m_X - f_{\sigma_X},\ m_X + f_{\sigma_X})$. Literature suggests that a suitable value for f is in the range of 1–3. Utilizing the chosen sample points yields

$2n + 1$ values of the function $g(X)$. Subsequently, the unknown coefficients in the response surface function can be calculated. Once the response surface function is obtained, approximate values for the design check points X_D on the limit state surface can be calculated.

The second step involves selecting a new center point, X_M, which can be chosen from the line connecting the mean point m_X and X_D, while ensuring the validity of the limit equation, $g(X) = 0$, namely:

$$X_M = m_X + (X_D - m_X)\frac{g(m_X)}{g(m_X) - g(X_D)} \tag{2.9}$$

This selection of a new center point aims to ensure that the original limit state surface's information is encompassed by the chosen sample points as much as possible.

The third step involves considering X_M as the center point to select a new set of sample points and then repeating the process from the first step. This will yield numerical values for the design check points on the limit state surface and related reliability indicators. The entire procedure requires solving for the values of $4n + 3$ functions $g(X)$. Figure 2.1 illustrates the schematic diagram of the entire approximation process.

However, in traditional quadratic response surface methods, when computing nonlinear functional functions, there are often issues with convergence failure and significant errors. Many scholars have proposed improved response surface methods, as well as intelligent response surface methods like Kriging (Luo et al. 2012), Artificial Neural Networks (ANN) (Cho 2009), Radial Basis Functions (RBF) (Deng et al. 2005), and Support Vector Machines (SVM) (Zhao 2008), to enhance the accuracy of structural probability reliability analysis. However, this also increases the complexity of the solution process. Additionally, for highly nonlinear structures such as high earth-rock dams, the effectiveness of sample training for these intelligent methods might not be ideal, and the improvement in accuracy for failure probability estimation might not be significant. It is important to note that first-order second-moment methods and response surface methods are generally used for the probability analysis of static systems. They are less frequently used in dynamic systems or are limited to

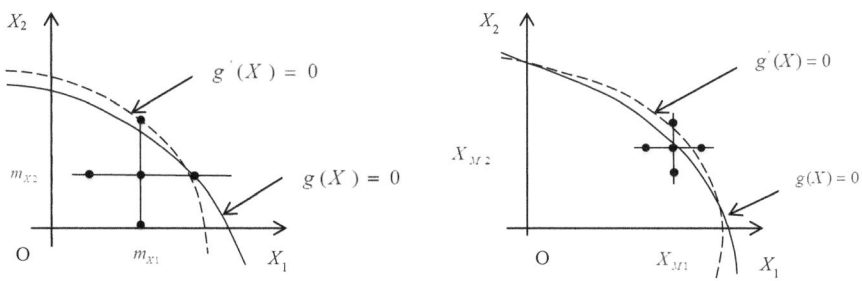

Fig. 2.1 The approximation process of quadratic response surface method

simple structural probability analysis. Moreover, they cannot adequately capture the stochastic dynamic response process.

In conclusion, the first-order second-moment method, although conceptually simple and computationally convenient, often lacks accuracy and struggles with convergence for complex problems. It may not guarantee precision and is unsuitable for solving engineering structural probability with implicit functions or complex state functions. The Monte Carlo method is versatile, but it's computationally inefficient and challenging to apply to complex real-world engineering problems. However, it can be used to validate the accuracy of other probability analysis methods. The response surface method is primarily employed for solving probability analysis problems with implicit state functions, but it has drawbacks such as convergence failures and significant errors. Some of its improvement methods often suffer from suboptimal sample training results and increased complexity in the solution process.

It is important to emphasize that the first-order second-moment method and the response surface method are generally used for the probability analysis of static systems. They are less frequently used in dynamic systems or are limited to simple structural probability analysis, and they cannot capture the stochastic dynamic response process. In recent years, the probability density evolution method based on stochastic vibration theory has made significant contributions to the field of structural probability analysis. It decouples the solution from complex state functions by solving physical equations and probability density evolution equations. It can incorporate all probability information about structural response and has been successfully applied to structural probability analysis. This approach has played a prominent role in the development and application of structural reliability probability theory.

2.3   Generalized Probability Density Evolution Method

In the 1940s, American scholar Housner (1947) conducted relevant research on the stochastic nature of earthquake ground motion, drawing the attention of seismic researchers from various countries. This marked the beginning of research into stochastic dynamic analysis of earthquakes, including the study of random vibration theory for structural earthquake responses. It found widespread applications in aerospace, mechanical, civil, bridge, and marine engineering fields. Random vibration involves treating input ground motion and resulting structural responses as random processes. Applying stochastic dynamic theory yields statistical characteristics of structural responses, enabling the estimation of failure probabilities. Over time, scholars both domestically and internationally have conducted extensive research based on power spectral analysis, moment evolution, and FPK equation. The study of linear systems within stochastic vibration theory has gradually matured, with improved computational efficiency, and it has been extensively applied in engineering. For instance, Zhu (1993) derived the steady-state solution of the FPK equation based on Hamilton's theory, Lin and Zhong. (1998) proposed the pseudo excitation method, and Fujimura and Kiureghian (2007) introduced the truncated equivalent linear method. Several stationary random earthquake motion models, such as the K-T model (Kanai 1957), C-P model (Clough 1993), S-O model (Ou and Niu 1990), and D-C model (Du and Chen 1994), have also been proposed or developed, yielding satisfactory results.

The aforementioned methods only address the randomness of structural loads. When considering the randomness of structural parameters, the classical stochastic vibration theory encounters significant challenges. On the other hand, earthquakes are typically composed of initiation, main shock, and decay phases. Strictly speaking, they should be considered non-stationary excitation processes. However, extensive research indicates that there is a distinct difference in seismic response between non-stationary and stationary stochastic ground motion models for the same nonlinear structure. Stationary stochastic ground motion models often underestimate structural dynamic responses. Furthermore, for highly nonlinear structures like earth-rock dams under seismic loads, traditional response spectrum analysis based on power spectral models is no longer applicable. Instead, refined simulation of random dynamic responses using seismic acceleration time-history analysis is needed.

Therefore, there is a need for the development and adoption of stochastic dynamic analysis methods tailored to highly nonlinear structures and non-stationary seismic motion models. Currently, some scholars have proposed methods such as stochastic simulation, random perturbation methods, and orthogonal polynomial expansion theory to solve the stochastic dynamic responses of nonlinear structures. However, these methods only yield approximations of second-order statistics and fail to provide complete probability information about structural responses (Liu 2013). For dynamic response analysis of structures, it's essential to consider the influence of various stochastic factors as comprehensively as possible. The stochastic vibration analysis of structures should encompass random dynamic responses and ultimately serve structural probability analysis.

Since 2003, Li and Chen from Tongji University (2003, 2010, 2017) have developed the concept of probability density evolution based on the fundamental idea of probability density evolution, known as the Generalized Probability Density Evolution Method (GPDEM). Starting from the state equation, this method utilizes the stochastic event characterization based on the principle of probability conservation to derive a decoupled generalized probability density evolution equation. It combines techniques such as probabilistic space point selection, deterministic structural numerical analysis, and finite difference methods to perform nonlinear stochastic vibration analysis of structural responses and dynamic reliability probability analysis. Moreover, the probability density function encompasses all stochastic factors in the system, laying a solid foundation for studying reliability probability analysis and uncertainty propagation in complex structures. This approach provides a more accurate description of structural dynamic behavior and has yielded positive results in seismic safety analysis of large and complex structures such as bridges, dams, and aqueducts (Liu et al. 2013, 2014; Liu and Fang 2012).

2.3.1 The Generalized Probability Density Evolution Equation

As is well known, external excitations, system parameters, and initial conditions of structural dynamic responses are all characterized by randomness. Structural dynamic responses can be regarded as stochastic processes, and their statistical probability characteristics are entirely determined by the aforementioned sources of randomness. Therefore, based on the principle of probability conservation, by establishing the probability density evolution rules from the source random factors to the target random responses, it is possible to solve the stochastic dynamic processes and reliability probabilities of complex nonlinear structures. This allows for a comprehensive understanding of their seismic performance under different levels of earthquake excitation.

Based on the knowledge of structural dynamics, the equation of motion for an n-degree-of-freedom system under dynamic loading excitation can be represented as follows:

$$\overline{\mathbf{M}}(\mathbf{\Theta})\ddot{\mathbf{X}}(t) + \mathbf{C}(\mathbf{\Theta})\dot{\mathbf{X}}(t) + \mathbf{K}(\mathbf{\Theta})\mathbf{X}(t) = -\overline{\mathbf{M}}\ddot{\mathbf{X}}_{\mathrm{g}}(\mathbf{\Theta}, t) \tag{2.10}$$

where, $\overline{M}$ $\mathbf{C}$, and $\mathbf{K}$ represent the effective mass, damping, and stiffness matrices of the structure, with their fundamental parameters possibly exhibiting randomness; $\ddot{\mathbf{X}}(t)$, $\dot{\mathbf{X}}(t)$ and $\mathbf{X}(t)$ are the acceleration, velocity, and displacement vectors of the structural response, respectively; $\ddot{\mathbf{X}}_{\mathrm{g}}(\mathbf{\Theta}, t)$ is the stochastic dynamic excitation process; $\mathbf{\Theta}$ is the random vector within the entire system, and the solution of Eq. (3.1) uniquely and continuously depends on $\mathbf{\Theta}$. For convenience, the solution of Eq. (2.10) can be written as:

$$\mathbf{X}(t) = \mathbf{H}(\mathbf{\Theta}, t) \tag{2.11}$$

where, $\mathbf{H} = (H_1, H_2, \ldots, H_{\mathrm{n}})^{\mathrm{T}}$. The velocity time history can be expressed as:

$$\ddot{\mathbf{X}}(t) = \mathbf{h}(\mathbf{\Theta}, t) \tag{2.12}$$

where, $\mathbf{h} = (h_1, h_2, \ldots, h_{\mathrm{n}})^{\mathrm{T}}$.

Therefore, more generally, any physical quantity of response in the structural system, such as deformation, displacement, velocity, acceleration, and stress, uniquely and continuously depend on $\mathbf{\Theta}$. Denote the interested physical quantity as $\mathbf{Z} = (Z_1, Z_2, \ldots, Z_{\mathrm{m}})^{\mathrm{T}}$, then:

$$\mathbf{Z}(t) = \mathbf{H}_Z(\mathbf{\Theta}, t) \tag{2.13}$$

$$\dot{\mathbf{Z}}(t) = \mathbf{h}_Z(\mathbf{\Theta}, t) \tag{2.14}$$

where, $\mathbf{h} = (h_{z,1}, h_{z,2}, ..., h_{Z,m})^{T}$.

Clearly, Eq. (2.13) can also be regarded as a stochastic dynamic system, where the source random factors are entirely described by $\boldsymbol{\Theta}$. Considering the augmented system formed by $(\mathbf{Z}, \boldsymbol{\Theta})$, since all the probabilistic factors are encompassed, it represents a conservative probabilistic system. The joint probability density function (PDF) of $(\mathbf{Z}, \boldsymbol{\Theta})$ can be denoted as $p_{Z\Theta}(z, \boldsymbol{\theta}, t)$. According to the principle of probability conservation (Chen and Li 2009), the following equation can be obtained:

$$\frac{D}{Dt} \int_{\Omega_t \times \Omega_{\Theta}} p_{Z\Theta}(z, \boldsymbol{\theta}, t) dz d\boldsymbol{\theta} = 0 \tag{2.15}$$

After undergoing certain mathematical manipulations (El Hami and Radi 2016), the above equation can be transformed into:

$$\frac{\partial p_{z\Theta}(z, \boldsymbol{\theta}, t)}{\partial t} + \sum_{l=1}^{m} h_{z,l}(\boldsymbol{\theta}, t) \frac{\partial p_{z\Theta}(z, \boldsymbol{\theta}, t)}{\partial z_l} = 0 \tag{2.16}$$

Combining Eq. (2.14), a more explicit conclusion can be drawn:

$$\frac{\partial p_{z\Theta}(z, \boldsymbol{\theta}, t)}{\partial t} + \sum_{l=1}^{m} \dot{Z}_1(\boldsymbol{\theta}, t) \frac{\partial p_{z\Theta}(z, \boldsymbol{\theta}, t)}{\partial z_l} = 0 \tag{2.17}$$

Therefore, the joint probability density function of $Z(t)$, denoted as $p_Z(z, t)$, is:

$$p_Z(z, t) = \int_{\Omega_{\Theta}} p_{z\Theta}(z, \boldsymbol{\theta}, t) d\boldsymbol{\theta} \tag{2.18}$$

When considering only a specific response physical quantity, Eq. (2.16) degenerates into a one-dimensional partial differential equation, namely:

$$\frac{\partial p_Z(z, \boldsymbol{\theta}, t)}{\partial t} + \dot{Z}(\boldsymbol{\theta}, t) \frac{\partial p_{z\Theta}(z, \boldsymbol{\theta}, t)}{\partial z} = 0 \tag{2.19}$$

The dimensionality m of Eq. (2.16) is independent of the original physical system's degree of freedom n. Regardless of whether the source random factors originate from initial conditions, structural parameters, or external excitations, the governing equation takes the form of Eq. (2.16). Therefore, Eq. (2.16) is referred to as the Generalized Probability Density Evolution Equation, sometimes simply referred to as the Generalized Density Evolution Equation (GDEE). From the above derivation, it is evident that the physical law revealed by the GDEE indicates that, during the evolution process of a general dynamic system, the rate of change of the joint probability density function distribution of the generalized displacement (which can represent actual displacement, stress, deformation, etc.) with respect to time is proportional to the rate of change of the generalized displacement. The proportionality coefficient is determined by the instantaneous generalized velocity, implying that the process of probability density evolution follows strict physical laws. The initial and boundary conditions for Eq. (2.16) are as follows:

$$p_{Z\Theta}(z, \ \boldsymbol{\theta}, \ t)|_{t=t_0} = \delta(z - z_0)p_{\Theta}(\boldsymbol{\theta}) \tag{2.20}$$

$$p_{Z\Theta}(z, \ \boldsymbol{\theta}, \ t)|_{z_j \to \pm\infty} = 0, \quad j = 1, \ 2, \ \dots, \ m \tag{2.21}$$

where z_0 represents the deterministic initial value. Hence, solving the GDEE is a combination of solving physical equations and solving the Probability Density Evolution Equation. For some simpler problems, analytical solutions can be obtained using methods like characteristic lines. However, for the majority of engineering practical problems, numerical solutions are needed. From Eq. (2.17), it's apparent that this is a linear partial differential equation. To obtain a numerical solution for this partial differential equation, the coefficients of the equation need to be obtained, which are the derivatives of the physical quantity under consideration when $\{\boldsymbol{\Theta} = \boldsymbol{\theta}\}$. These derivatives can be acquired from the solutions of Eqs. (2.10) and (2.14). Therefore, the numerical implementation of the probability density evolution theory can follow the steps below, as shown in Fig. 2.2. The process of solving the GDEE can be divided into the following four steps:

(1) Probabilistic space point selection and assignment of probabilities: In the distribution space $\boldsymbol{\Omega}_{\boldsymbol{\Theta}}$ of the fundamental random variable $\boldsymbol{\Theta}$, a set of discrete representative points $\boldsymbol{\theta}_q$ ($q = 1, 2, \dots, n_{sel}$) is selected using methods like number theory, quasi-Monte Carlo, spherical simplex, and GF-deviation, among others. Here, n_{sel} represents the number of discrete representative points. Meanwhile, the assigned probability $P_q = \int_{V_q} p_{\Theta}(\boldsymbol{\theta})d\boldsymbol{\theta}$ for each representative point is determined. V_q denotes the representative volume.

(2) Deterministic dynamic system solution: For each given $\boldsymbol{\Theta} = \boldsymbol{\theta}_q$, solve the physical Eqs. (2.10) and (2.14) to obtain the time derivatives (velocities) $\dot{Z}_j(\boldsymbol{\theta}_q, t_m)$ ($j = 1, 2, \dots, m$) of the desired physical quantities. It's worth noting that when the random parameters are determined through probabilistic space point selection, the differential equations of the stochastic dynamic system are transformed into a set of deterministic dynamic equations. For engineering structures, these equations can be solved using various numerical simulation methods such as finite element method and finite difference method.

(3) Solving the GDEE: After the first step of discrete representative point selection and assignment of probabilities, the obtained GDEE is shown below:

$$\frac{\partial p_{Z\Theta}(z, \ \boldsymbol{\theta}_q, \ t)}{\partial t} + \sum_{j=1}^{m} \dot{Z}_j(\boldsymbol{\theta}_q, \ t)\frac{\partial p_{Z\Theta}(z, \ \boldsymbol{\theta}_q, \ t)}{\partial z_j} = 0 \tag{2.22}$$

The corresponding initial conditions become:

$$p_{\mathbf{Z}\Theta}(z,\ \boldsymbol{\theta}_q,\ t)|_{t=t_0} = \delta(z-z_0)P_q \qquad (2.23)$$

Substituting the expression, $\dot{Z}_j(\boldsymbol{\theta}_q,\ t_m)$, obtained in the second step into Eqs. (2.22) and (2.23), and utilizing certain numerical methods such as finite difference method, the partial differential equation can be solved to obtain its numerical solution.

(4) Cumulative summation: By cumulatively summing up all the discrete numerical solutions, $p_{\mathbf{Z}\Theta}(z,\ \boldsymbol{\theta}_q,\ t)$, obtained above, the numerical solution of $p_{\mathbf{Z}}(z,\ t)$ can be obtained.

$$p_{\mathbf{Z}}(z,\ t) = \sum_{q=1}^{n_{sel}} p_{\mathbf{Z}\Theta}(z,\ \boldsymbol{\theta}_q,\ t) \qquad (2.24)$$

As seen, the solution of the Generalized Probability Density Evolution process is essentially based on the principle of probability conservation. It transforms the stochastic dynamic system into a series of deterministic physical equations that

Fig. 2.2 Solving flowchart of GPDEM

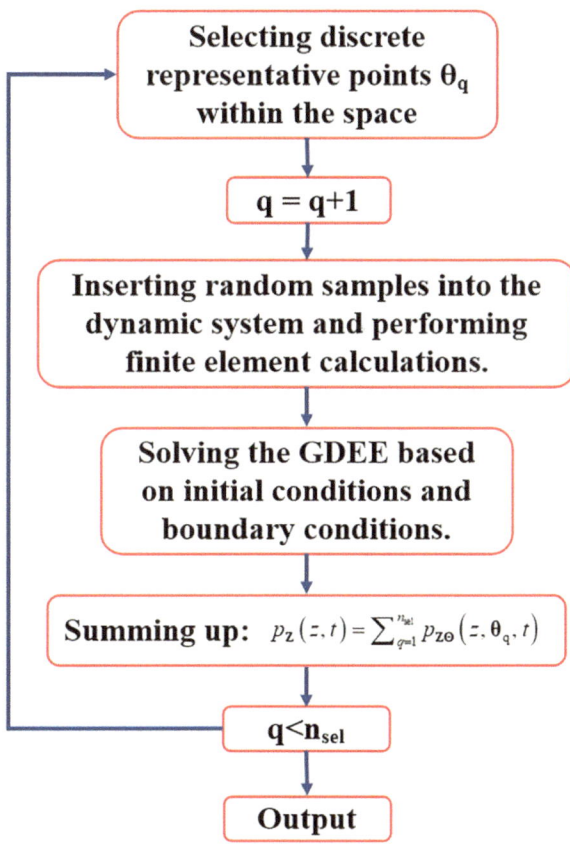

possess inherent probabilistic connections. Through the GDEE, it acquires the probabilistic information about the system's physical state. This process combines the solutions of a series of deterministic dynamic systems and the solution of the GDEE. This exactly embodies the fundamental idea that the evolution of probability density functions depends on the evolution mechanism of the physical system's state. Moreover, the solution of the dynamic system and the solution of the GDEE are decoupled, avoiding the need for repetitive iterations, or solving implicit functions as in traditional reliability probability methods. This approach demonstrates good applicability for complex nonlinear dynamic systems.

2.3.2   The Selection Method of Discrete Representative Points in Probability Space

The selection of discrete representative points in probabilistic space is one of the crucial techniques in the application of the probability density evolution method. Assuming a multidimensional probabilistic space Ω_Θ contains a set of discrete representative points $\boldsymbol{\theta}_q = (\theta_{1,q}, \theta_{2,q}, ..., \theta_{s,q})$, where $q = 1, 2, ..., n_{sel}$. For each representative point, if we take the Voronoi region as the representative volume (as shown in Fig. 2.3), the probability within this representative volume corresponds to the assigned probability of that representative point:

$$P_q = \Pr\{\boldsymbol{\Theta} \in V_q\} = \int_{V_q} P_\Theta(\boldsymbol{\theta})\mathrm{d}\boldsymbol{\theta} \quad q = 1, 2, \ldots, n_{sel} \tag{2.25}$$

$$\lim_{r_{cv} \to 0} \tilde{p}_\Theta(\boldsymbol{\theta}) = p_\Theta(\boldsymbol{\theta}) \tag{2.26}$$

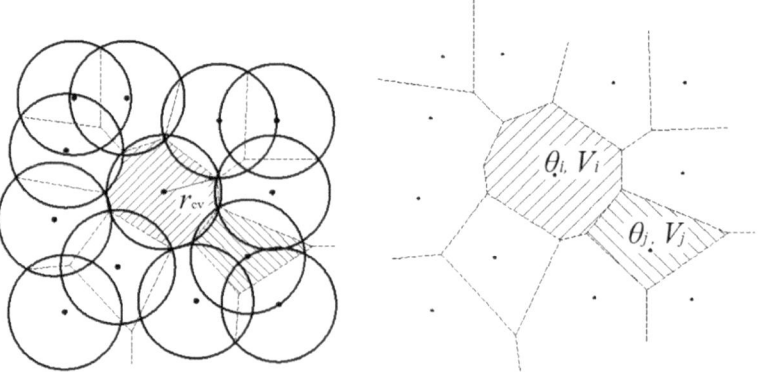

Fig. 2.3 Partition of probability space

where r_{cv} is the covering radius of the point set $\mathcal{P}_{sel} =$ $\{\boldsymbol{\theta}_q = (\theta_{1,q}, \theta_{2,q}, \cdots, \theta_{s,q}); \; q = 1, 2, \cdots, n_{sel}\}$. It can be easily deduced that:

$$
\int_{\Omega_\Theta} p_\Theta(\boldsymbol{\theta}) d\boldsymbol{\theta} = \int_{\Omega_\Theta} \tilde{p}_\Theta(\boldsymbol{\theta}) d\boldsymbol{\theta} = \sum_{q=1}^{n_{sel}} P_q = \sum_{q=1}^{n_{sel}} \int_{V_q} p_\Theta(\boldsymbol{\theta}) d\boldsymbol{\theta}
$$

$$
= \int_{U_{q=1}^{n_{sel}} V_q} p_\Theta(\boldsymbol{\theta}) d\boldsymbol{\theta} = 1 \tag{2.27}
$$

This is the compatibility condition of assigned probabilities. In the probabilistic space, the selection of discrete fundamental point sets should have minimal deviations; based on the probability distribution type, suitable transformations are applied to the fundamental point set to minimize the F-deviation. The formula for calculating the F-deviation is:

$$
\mathcal{D}_{\mathcal{F}}(n, \mathcal{P}) = \sup_{x \in R^s} |\mathcal{F}_n(x) - \mathcal{F}(x)| \tag{2.28}
$$

$$
\mathcal{F}_n(x) = \frac{1}{n} \sum_{q=1}^{n} I\{x_q \leq x\} \tag{2.29}
$$

where $\mathcal{F}(x)$ is the joint distribution function and $\mathcal{F}_n(x)$ is the empirical distribution function of the representative point set. $I\{\cdot\}$ is the indicator function. Considering the assigned probabilities of the representative point set, the F-deviation in Eq. (2.29) can be further modified as:

$$
\mathcal{F}_n(x) = \sum_{q=1}^{n} P_q \cdot I\{x_q \leq x\} \tag{2.30}
$$

Typically, discrete representative point selection can be achieved using methods such as the sphere-cutting method, lattice point method, number-theoretic methods, and the GF-deviation-based optimization method, significantly reducing the number of selected points. Here, we will briefly introduce the number-theoretic point selection method and the GF-deviation-based point set optimization method used in this paper.

(1) Number-Theoretic Method: Let there be a vector composed of a set of integers $(n, Q_1, Q_2, ..., Q_s)$, and according to the following formula, a set of points can be generated in an s-dimensional space:

$$
\hat{x}_{j,k} = (2kQ_j - 1) \bmod (2n); \quad j = 1, 2, \ldots, s; \; k = 1, 2, \ldots, n
$$

$$
x_{j,k} = \frac{\hat{x}_{j,k}}{2n} \tag{2.31}
$$

where mod$(\cdot)$ represents the remainder after division. The above equation is equivalent to:

$$x_{j,k} = \frac{2kQ_j - 1}{2n} - \text{int}\left(\frac{2kQ_j - 1}{2n}\right); \quad j = 1, 2, \ldots, s; \ k = 1, 2, \ldots, n \tag{2.32}$$

where int$(\cdot)$ represents the integer part of the expression inside the brackets, and n is the total number of points in the number-theoretic point set, $\mathcal{P}_{\text{NTM}}$.

Clearly, the numbers in Eqs. (2.31) and (2.32) satisfy:

$$0 < x_{j, k} < 1, j = 1, 2, \ldots, s; \ k = 1, 2, \ldots, n \tag{2.33}$$

By appropriately selecting the integer vector $(n, Q_1, Q_2, \ldots, Q_s)$, it is possible to generate a point set using Eq. (2.33) with a smaller deviation, such as the point set generation method proposed by Hua and Wang (1981).

(2) GF-Deviation Method: Chen et al. (2016) developed a GF-deviation minimization point set optimization method for non-uniform and non-normal multidimensional distributions. It is mainly achieved in two steps. Step 1: Generate an initial point set using the Sobol sequence (Radović et al. 1996), then rearrange the point set to minimize the GF-deviation. The initial point sets $\mathbf{x}_q = (x_{q,1}, x_{q,2}, \ldots, x_{q,i})$, obtained from Sobol point sets $\mathbf{u}_q = (u_{q,1}, u_{q,2}, \ldots, u_{q,i})$ $(q = 1, 2, \ldots, n;$ i represents the ith random variable) over the unit hypercube, are given by the following expressions:

$$x_{\text{m},i} = F_i^{-1}\left(u_{\text{m},i}\right) \tag{2.34}$$

In the expression, $F_i^{-1}(\cdot)$ is the inverse cumulative distribution function of the ith random variable. Step 2: Transform each random variable so that the assigned probabilities of the sets of n points become mutually close. The assigned probabilities $x_q^* = (x_{q, 1}^*, x_{q, 2}^*, \ldots, x_{q,i}^*)$ points are estimated:

$$x_{m, i}^* = F_i^{-1}\left(\sum_{q=1}^{n} \frac{1}{n} \cdot I\{x_{q, i} < x_{m, i}\} + \frac{1}{2} \cdot \frac{1}{n}\right) \tag{2.35}$$

To reduce the GF bias, the following transformation is applied:

$$x_{m,i}^{**} = F_i^{-1}\left(\sum_{q=1}^{n} p_q \cdot I\{x_{q, i}^* < x_{m, i}^*\} + \frac{1}{2} \cdot p_m\right) \tag{2.36}$$

Finally, $x_q^{**} = (x_{q, 1}^{**}, x_{q, 2}^{**}, \ldots, x_{q, i}^{**})$ represents the representative set of points used.

2.3.3 Numerical Solution Methods

In some simpler cases, analytical solutions for the probability density evolution equation can be obtained. However, for complex multi-degree-of-freedom structural solutions, numerical methods are typically employed, such as various forms of finite difference methods and finite element methods. Here, we briefly introduce the Total Variation Diminishing (TVD) finite difference method and the Streamline Upwind Petrov–Galerkin (SUPG) finite element method used in this paper.

(1) Total Variation Diminishing (TVD) finite difference method.

To solve the GDEE, the Total Variation Diminishing (TVD) finite difference method is often employed. Equation (2.21) can be discretized as follows:

$$p_j^{(k+1)} = p_j^{(k)} - \frac{1}{2}(\lambda a - |\lambda a|)\Delta p_{j+\frac{1}{2}}^{(k)} - \frac{1}{2}(\lambda a + |\lambda a|)p_{j-\frac{1}{2}}^{(k)}$$
$$- \frac{1}{2}(|\lambda a| - \lambda^2 a^2)\left(\psi_{j+\frac{1}{2}}\Delta p_{j+\frac{1}{2}}^{(k)} - \psi_{j-\frac{1}{2}}\Delta p_{j-\frac{1}{2}}^{(k)}\right) \tag{2.37}$$

where:

$$r_{j+\frac{1}{2}}^+ = \frac{\Delta p_{j+\frac{3}{2}}^{(k)}}{\Delta p_{j+\frac{1}{2}}^{(k)}} = \frac{\Delta p_{j+2}^{(k)} - \Delta p_{j+1}^{(k)}}{\Delta p_{j+1}^{(k)} - \Delta p_j^{(k)}}, r_{j+\frac{1}{2}}^- = \frac{\Delta p_{j-\frac{1}{2}}^{(k)}}{\Delta p_{j+\frac{1}{2}}^{(k)}} = \frac{\Delta p_j^{(k)} - \Delta p_{j-1}^{(k)}}{\Delta p_{j+1}^{(k)} - \Delta p_j^{(k)}} \tag{2.38}$$

Introducing the expression for flux limiter:

$$\psi_{j+\frac{1}{2}}(r_{j+\frac{1}{2}}^+, r_{j+\frac{1}{2}}^-) = u(-a)\psi_0(r_{j+\frac{1}{2}}^+) + u(a)\psi_0(r_{j+\frac{1}{2}}^-) \tag{2.39}$$

In the equation, $u(\cdot)$ represents the Heaviside function, and $\psi_0(r) = \max(0, \min(2r, 1), \min(r, 2))$. Research indicates that the difference scheme (Eq. 2.38) possesses the TVD (Total Variation Diminishing) property:

$$\mathrm{TV}[p(\cdot, t_2)] \leq \mathrm{TV}[p(\cdot, t_1)] \leq \mathrm{TV}[p(\cdot, t_0)], \quad t_2 > t_1 > t_0, \tag{2.40}$$

where the total variation is:

$$\mathrm{TV}[p(\cdot, t)] = \int_{-\infty}^{\infty} |\frac{\partial p(x, t)}{\partial x}| dx \tag{2.41}$$

For the case of discrete curves:

$$\mathrm{TV}(p_.^{(k)}) = \sum_{j=-\infty}^{\infty} |p_{j+1}^k - p_j^k| \tag{2.42}$$

(2) Streamline Upwind Petrov–Galerkin (SUPG) finite element method.

The SUPG finite element method possesses better convergence properties and can obtain the probability density function at boundary locations more effectively. Based on similarity, Eq. (2.21) can be expressed as follows:

$$\mathbf{w} \cdot \nabla p = 0 \tag{2.43}$$

where $\mathbf{w} = (1, \alpha(x))$ represents the velocity field, $\alpha(x)$ is the flux. and $\Omega = \Omega_u \times \Omega_x \subset R^2$ denotes the computational domain of the finite element partition, where $e = 1, 2, \ldots, N_{el}$. As the domain is two-dimensional, we can discretize it using quadrilateral elements.

According to reference (Elman et al. 2014), we can define the trial solution space as, where $V = \{p | p \in \mathcal{H}^1(\Omega), p = g \ on \ \partial\Omega\}$ is the Sobolev space. We also define the space: $V_0 = \{\Psi | \Psi \in \mathcal{H}^1(\Omega), \Psi = 0 \ on \ \partial\Omega\}$. The fundamental idea of Petrov–Galerkin approximation is to specify a weak formulation where the space of test (weighting) functions is different from that of the trial solution. More specifically, the test space is spanned by functions of the form given in Eq. (2.44):

$$\widetilde{\psi} = \psi + \tau \mathbf{w} \nabla \psi \tag{2.44}$$

where $\Psi \in V_0$ is the Galerkin-type weighting function, τ is a coefficient, and element e is given by the following expression:

$$\tau_e = \frac{a\lambda_e}{2|\mathbf{w}_e|} \tag{2.45}$$

In the above expression, e represents the characteristic length of element $\tau_e = \min \frac{\lambda_x}{\cos\vartheta}, \frac{\lambda_u}{\sin\vartheta}$, λ_x and λ_u denote the lengths of the rectangle in the x and u directions, respectively, where $\vartheta = \arctan\left(\left|\frac{\omega_u}{\omega_x}\right|\right)$. The following equations will be denoted by subscript h to indicate the discrete finite element problem. The weak form of Eq. (2.43) is:

$$\int_\Omega \psi^h \mathbf{w} \nabla p^h d\Omega + \sum_{e=1}^{n_{el}} \int_{\Omega_e} \tau \mathbf{w} \nabla \psi^h \mathbf{w} \nabla p^h d\Omega_e = 0 \tag{2.46}$$

To further enhance the accuracy of the method, we incorporate the approach proposed in (Hughes et al. 1986) into Eq. (2.46), resulting in:

$$\int_\Omega \psi^h \mathbf{w} \nabla p^h d\Omega + \sum_{e=1}^{n_{el}} \int_{\Omega_e} \tau_1 \mathbf{w} \nabla \psi^h \mathbf{w} \nabla p^h d\Omega_e$$

$$+ \sum_{e=1}^{n_{el}} \int_{\Omega_e} \tau_2 \mathbf{w} \nabla \psi^h \mathbf{w}_{\parallel} \nabla p^h d\Omega_e = 0 \qquad (2.47)$$

Here, $\mathbf{w}_{\parallel}$ represents the projection of $\mathbf{w}$ onto the values on ∇p^h:

$$\mathbf{w}_{\parallel} = \begin{cases} \frac{\mathbf{w} \nabla p^h}{|\nabla p^h|_2^2}, & \text{if } \nabla p^h \neq 0 \\ 0 & \text{if } \nabla p^h = 0 \end{cases} \qquad (2.48)$$

2.4 Non-Stationary Stochastic Seismic Motion Model

Seismic motion significantly influences the seismic response of structures, exhibiting pronounced randomness in both intensity and frequency. However, most of the current research is based on deterministic analysis methods. Therefore, it is essential to employ stochastic dynamic theory based on dynamic time history analysis to explore the seismic response of structures under stochastic seismic excitations. Utilizing the numerical solution process of the generalized probability density evolution method, the probability density evolution equation discretizes the probability space constituted by stochastic factors. As a result, the number of acceleration time history samples is determined by the quantity of discretized representative points. It is evident that the non-stationary stochastic seismic motion model forms the foundation for analyzing the random seismic response of engineering structures and seismic reliability probability using the generalized probability density evolution method.

2.4.1 Improved Clough-Penzien Power Spectral Model

Large-scale structures, especially earth-rock dams, are quite intricate. Under seismic actions, they exhibit nonlinear and even strongly nonlinear effects. Nonlinear dynamic analysis is necessary. The traditional response spectrum method is no longer applicable, and seismic acceleration time history analysis must be employed to comprehensively understand the seismic response process. Conventional artificial synthesis methods for stochastic seismic acceleration time histories seldom consider the influence of non-stationary characteristics of seismic motion, which is unreasonable. Therefore, it is appropriate to use frequency-dependent non-stationary power spectral models to synthesize seismic acceleration time histories. In recent years, some researchers have introduced frequency-dependent non-stationary power spectral models into the field of hydraulic structures to study their effects on dam bodies. However, this area has been less explored in the context of earth-rock dams, especially high-panel block dams. Nonetheless, based on the characteristics of high-panel block dams, the study of non-stationary seismic stochastic dynamic excitation is one of the effective paths for the performance-based seismic safety assessment of high-panel block dams. In 1996, Deodatis proposed an evolving power spectral model for fully non-stationary seismic acceleration time histories based on the stationary Clough-Penzien power spectral model. In 2011, Cacciola and Deodatis improved the model. The bilateral evolving power spectral density function can be constructed using the following equation:

$$S_{\ddot{X}_g}(t,\ \omega) = A^2(t) \frac{\omega_g^4(t) + 4\xi_g^2(t)\omega_g^2(t)\omega^2}{[\omega^2 - \omega_g^2(t)]^2 + 4\xi_g^2(t)\omega_g^2(t)\omega^2}$$
$$\cdot \frac{\omega^4}{[\omega^2 - \omega_f^2(t)]^2 + 4\xi_f^2(t)\omega_f^2(t)\omega^2} S_0(t) \qquad (2.49)$$

In the equation: $A(t)$ is the intensity modulation function, recommended to be taken as (Cacciola and Deodatis 2011):

$$A(t) = \left[\frac{t}{c} \exp\left(1 - \frac{t}{c}\right)\right]^d \qquad (2.50)$$

In the equation, c represents the time at which the peak ground acceleration (PGA) of the seismic motion occurs, and in this paper, it is set to 4 s; d is a parameter controlling the shape of $A(t)$, and in this paper, it is set to 2. In the modulation function of the evolving power spectral density, the following parameters reflect its frequency-dependent non-stationary characteristics:

$$\omega_g(t) = \omega_0 - a\frac{t}{T}, \xi_g(t) = \xi_0 + b\frac{t}{T} \qquad (2.51)$$

$$\omega_f(t) = 0.1\omega_g(t),\ \xi_f(t) = \xi_g(t) \qquad (2.52)$$

In the above equation, ω_0 and ξ_0 are the initial circular frequency and initial damping ratio, which can be determined by site characteristics. In this paper, they are taken as 25 and 0.45 rad/s, respectively. a and b are parameters determined based on site characteristics and seismic category. Their values are 3.5 and 0.3 rad/s, respectively. T represents the duration of the seismic acceleration time history, and its value varies depending on the site. Typically, for Site Classes I_0, I_1, II, III and IV, the values are taken as 12s, 15s, 20s, 25s, and 30s respectively. The spectral parameters that reflect spectral intensity can be expressed as:

$$S_0(t) = \frac{\bar{a}_{max}^2}{\gamma^2 \pi \omega_g(t)[2\xi_g(t) + 1/(2\xi_g(t))]} \qquad (2.53)$$

where $\bar{a}_{max}$ is the mean of the peak ground acceleration (PGA) of the seismic motion, and γ is the equivalent peak factor, taken as Eq. (2.7).

2.4.2 Random Seismic Generation Based on Spectral Representation-Stochastic Process

Building upon the generalized Clough-Penzien power spectral model, this paper employs the concept of stochastic processes to achieve the generation of non-stationary seismic motion using a spectral representation based on hydraulic seismic design codes. By combining this approach with the generalized probability density evolution method, the paper conducts refined stochastic dynamic response and seismic reliability analysis for high-panel block dams. Typically, non-stationary seismic acceleration random processes with zero mean can be generated using the following formula (Ou and Wang 1998):

$$\ddot{X}_g(t) = \sum_{k=1}^{N} \sqrt{2 S_{\ddot{X}_g}(t, \omega_k) \Delta\omega} [\cos(\omega_k t) X_k + \sin(\omega_k t) Y_k] \qquad (2.54)$$

In the equation, $\omega_k = k\Delta\omega$, and $S_{\ddot{X}_g}$ is the bilateral evolving power spectral density function. At the frequency $\omega = 0$, it should satisfy:

$$S_{\ddot{X}_g}(t, \omega_0) = S_{\ddot{X}_g}(t, 0) = 0 \qquad (2.55)$$

In Eq. (2.54), $\{X_k, Y_k\}$ $(k = 1, 2, …, N)$ are standard orthogonal random variables that satisfy the following fundamental conditions:

$$E[X_k] = E[Y_k] = 0 \qquad (2.56)$$

$$E[X_j Y_k] = 0, E[X_j X_k] = E[Y_j Y_k] = \delta_{jk} \qquad (2.57)$$

In the equation, $E[\cdot]$ represents the mathematical expectation, and δ_{jk} is the Kronecker delta. The relative error of the mean for simulating non-stationary seismic acceleration processes can be expressed as:

$$\varepsilon(N) = 1 - \frac{\int_0^{\omega_u} \int_0^T S_{\ddot{X}_g}(t, \omega) dt d\omega}{\int_0^{\infty} \int_0^T S_{\ddot{X}_g}(t, \omega) dt d\omega} \qquad (2.58)$$

In the equation: $\omega_u = N\Delta\omega$ represents the truncation frequency, T is the duration of the non-stationary seismic acceleration process. Typically, the relative error of the mean $\varepsilon(N) \leq 1.0$ is less than or equal to 1.0. In this paper, $\Delta\omega$ is set to 0.15 rad/s, and the truncation term N is taken as 1600.

Then, based on the concept of stochastic processes, the random function expression of the standard orthogonal random vector $\{X_k, Y_k\}$ can be constructed. Assuming that any two sets of standard orthogonal random vectors $\overline{X}_n$ and $\overline{Y}_n$ ($n = 1, 2, ..., N$) are functions of two independent random variables Θ_1 and Θ_2, the random function can be denoted as:

$$\overline{X}_n = \cas(n\Theta_1) \quad \overline{Y}_n = \cas(n\Theta_2) \tag{2.59}$$

In the equation, $cas(\mathrm{x}) = \cos(\mathrm{x}) + \sin(\mathrm{x})$ is the Hartley orthogonal basis function. The fundamental random variables Θ_1 and Θ_2 are uniformly distributed and mutually independent in the interval $[0, 2\pi]$, which can usually be obtained using number-theoretical methods. After a certain deterministic mapping, $\{X_k, Y_k\}$ becomes the standard orthogonal basis random variable needed for Eq. (2.54) and is uniquely determined.

Taking Site I_1 as an example, an acceleration time history with a duration of 15s is generated. The discretized representative points $\Theta_{1,i}$ and $\Theta_{2,i}$ are obtained using number-theoretical methods. To reduce fitting errors, the evolving power spectral density can often be corrected using the following equation:

$$S_{\ddot{X}_g}(t, \omega)|_{m+1} = \begin{cases} S_{\ddot{X}_g}(t, \omega), & 0 < \omega \leq \omega_c \\ S_{\ddot{X}_g}(t, \omega)|_m \frac{RSA^T(\omega, \zeta)^2}{RSA^S(\omega, \zeta)^2|_m}, & \omega > \omega_c \end{cases} \tag{2.60}$$

$S_{\ddot{X}_g}(t, \omega)|_{m+1}$ and $S_{\ddot{X}_g}(t, \omega)|_m$ are the evolving power spectra obtained after the $m + 1$ and m iterations, respectively. $RSA^S(\omega, \zeta)^2|_m$ is the average response spectrum obtained in the m iteration. $RSA^T(\omega, \zeta)^2$ is the target spectrum, often based on seismic design codes. Here, $\omega = 2\pi/T_0$, T_0 is the natural period of vibration. ζ is the damping ratio, typically taken as 0.05 for earth-rock dams. ω_c is the cutoff frequency, which can be set as 1.57 rad/s. By substituting Eq. (2.60) into Eq. (2.54), a set of representative acceleration time histories can be generated. Generally, after a few iterations, the required accuracy can be achieved. To provide a detailed description of the differences between the generated acceleration time histories and the target values, the following relative error controls are commonly employed:

$$\varepsilon_m = \frac{1}{N_m} \cdot \sum_{k=1}^{N_m} \left| \frac{X(t_k) - \overline{X}(t_k)}{Y(t_k)} \right| \tag{2.61}$$

$$\varepsilon_s = \frac{1}{N_m} \cdot \sum_{k=1}^{N_m} \left| \frac{Y(t_k) - \overline{Y}(t_k)}{Y(t_k)} \right| \tag{2.62}$$

$$\varepsilon_r = \frac{1}{N_r} \cdot \sum_{k=1}^{N_r} \left| \frac{R(T_{0,k}) - \overline{R}(T_{0,k})}{R(T_{0,k})} \right| \tag{2.63}$$

In the equation, ε_m represents the average relative error of the mean acceleration time history; ε_s is the average relative error of the standard deviation of the acceleration time history; ε_r is the average relative error of the response spectrum mean. $X(t_k)$

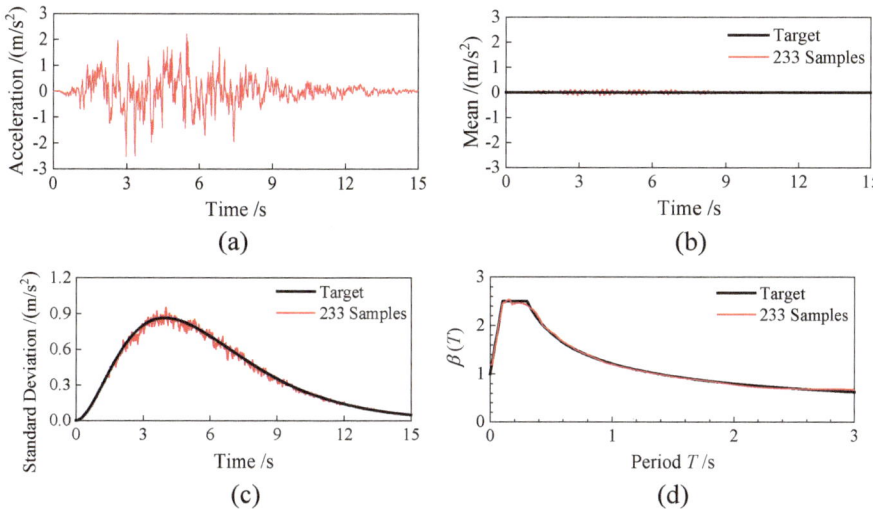

Fig. 2.4 Comparison of ground motion between generating samples and target, **a** typical sample acceleration; **b** average acceleration; **c** standard deviation of acceleration; **d** response spectrum

and $Y(t_k)$ are the mean and standard deviation of the target values at the kth control point. $\overline{X}(t_k)$ is the sample mean of the kth control point, and $\overline{Y}(t_k)$ is the sample standard deviation. $N_m = 1500$ is the number of time history discrete points. $R(T_0, k)$ is the response spectrum obtained based on seismic design codes, such as hydraulic seismic design codes. $\overline{R}(T_{0,k})$ is the response spectrum of the generated samples. $N_r = 400$ is the number of discrete points for natural period. When generating 233 sets of acceleration time histories, the resulting errors are 4.8% for ε_m, 4.2% for ε_s, and 3.4% for ε_r. Figure 2.4 provides a comparison between the generated seismic motion sample mean, standard deviation, and response spectrum, and the target values (with $\overline{a}_{max} = 2$). It can be observed that the fit is very good.

2.5 Dynamic Reliability Probability Analysis

In many cases, the statistics based on random dynamic response can describe the seismic response information of the structure and judge whether it is safe or not. However, based on a certain failure or failure criterion, it is more reasonable and more intuitive to give the reliability probability information of the system. It is also the main purpose of stochastic dynamics research to quantitatively evaluate the safety of structural systems from the perspective of failure probability, and it is also an important part of performance-based seismic safety evaluation of high concrete face rockfill dams. The dynamic reliability of engineering structures usually includes the reliability of first-passage failure and cumulative damage failure, which can be obtained by constructing a virtual stochastic process and solving the corresponding generalized probability density evolution equation.

For stochastic dynamical systems, within a given time interval $[0, T]$, the extreme value or cumulative value depends on the random vector $\mathbf{\Theta}$. Taking the extreme value distribution $X(t)$ as an example, the extreme value under seismic action can be expressed as:

$$\mathbf{Y_X} = \max \ (|\mathbf{H_X}(\mathbf{\Theta}, \ T)|, \ \mathbf{t} \in [0, \ \mathbf{T}]) \tag{2.64}$$

For a given $\mathbf{\Theta}$, $\mathbf{Y_X}$ exists and is unique, so there is

$$\mathbf{Y_X} = \mathbf{W_X}(\mathbf{\Theta}, T) \tag{2.65}$$

Therefore, a virtual stochastic process can be constructed:

$$\mathbf{Q_X}(\tau) = \mathbf{Y_\Theta}\tau = \mathbf{W_X}(\mathbf{\Theta}, \ T)\tau \tag{2.66}$$

where, τ is a virtual time parameter. Obviously there are:

$$\mathbf{Q_X}(\tau)|_{\tau=0} = 0, \mathbf{Y_X} = \mathbf{Q_X}(\tau)|_{\tau=1} \tag{2.67}$$

Derivation of Eq. (2.67) with respect to τ, then

$$\dot{\mathbf{Q}}_\mathbf{X} = \frac{\partial \mathbf{Q_X}}{\partial \tau} = \mathbf{W_X}(\mathbf{\Theta}, \ T) \tag{2.68}$$

Thus, $(\mathbf{Q}(\tau), \mathbf{\Theta})$ forms a conservative probability system, and the joint probability density equation $p_{Q\Theta}(q, \boldsymbol{\theta}, \boldsymbol{\tau})$ can be written as:

$$\frac{\partial p_{Q\Theta}(\mathbf{q}, \ \boldsymbol{\theta}, \ \tau)}{\partial \tau} + W(\boldsymbol{\theta}, \ \mathrm{T})\frac{\partial p_{Q\Theta}(\mathbf{q}, \ \boldsymbol{\theta}, \ \tau)}{\partial q} = 0 \tag{2.69}$$

Finally, the generalized probability density evolution equation method is used to solve the Eq. (2.69) to obtain the corresponding reliability or failure probability.

2.6 Verification and Application of the Examples

To validate the computational efficiency and accuracy of the Generalized Probability Density Evolution Method (GPDEM) for stochastic dynamic problems, especially in the stochastic analysis and probability assessment of nonlinear and complex engineering structures, this section will conduct verification in two aspects: seismic load randomness and parameter randomness. This will be achieved by utilizing nonstationary stochastic seismic ground motion models and the GF-bias method to generate acceleration time history samples and high-dimensional random parameter samples. These samples will be compared with analytical solutions, the Duffing oscillator, and multi-layered soil-rock slopes to obtain second-order statistical quantities and probability information. Finally, the applicability of the method for earth-rock dams will be primarily demonstrated through the verification on a panel-stacked rockfill dam.

2.6.1 Verification Based on Analytical Solution

This section verifies the accuracy and efficiency of GPDEM based on an undamped single-degree-of-freedom (SDOF) system. The equation of free vibration is as follows:

$$\ddot{u}(t) + \omega^2 u(t) = 0 \tag{2.70}$$

where the initial conditions are as follows:

$$u(t)|_{t=0} = x_0, \dot{u}(t)|_{t=0} = \dot{x}_0 \tag{2.71}$$

The natural frequency ω is a random parameter uniformly distributed in the interval $\left[\frac{5\pi}{4}, \frac{7\pi}{4}\right]$, with initial values of $x_0 = 1.0$m and $\dot{x}_0 = 0.0$m, respectively. When $\dot{x}_0 = 0.0$m, the displacement solution for the dynamic system Eq. (2.70) is:

$$X(t) = x_0 \cos(\omega t) \tag{2.72}$$

Clearly, $X(t)$ is a stochastic process. Ultimately, the probability density function (PDF) of the displacement and the second-order statistical time history (mean and standard deviation) can be obtained, as shown in Figs. 2.5 and 2.6. The excellent fit between the analytical solution and GPDEM demonstrates the high accuracy of this method.

2.6.2 Verification Based on Duffing Equation

To verify the accuracy and efficiency of employing the GPDEM method in solving the stochastic processes and reliability of nonlinear dynamic systems under random seismic effects, further investigation is conducted based on the typical nonlinear vibration system, the Duffing oscillator (Sekar and Narayanan 1994). The response of the Duffing oscillator under random seismic effects can be expressed as follows:

$$\ddot{x}(t) + 2\zeta_0 a_0 \dot{x}(t) + a_0^2 \left[x(t) + \mu x^3(t)\right] = -\ddot{U}_g(t) \tag{2.73}$$

Fig. 2.5 Comparison of
PDF between the GPDEM
and exact solution

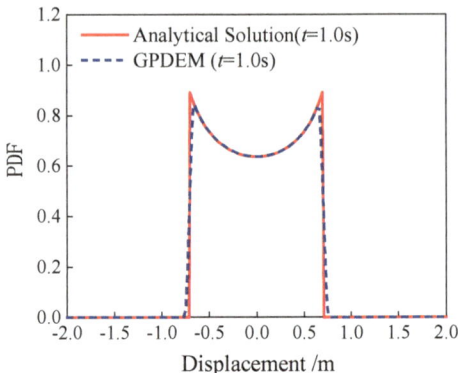

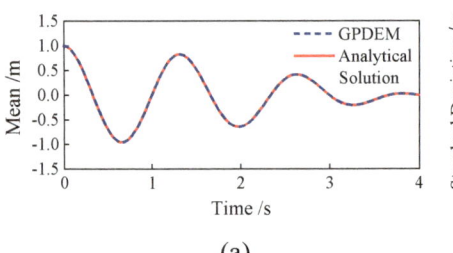

(a)

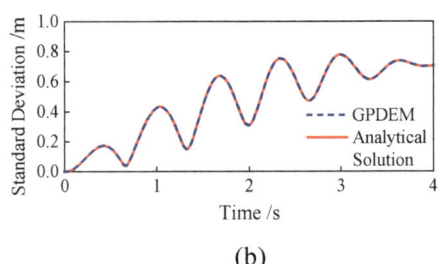

(b)

Fig. 2.6 Comparison of second-order statistical time-histories between the GPDEM and exact solution, **a** mean; **b** standard deviation

Here, where $a_0 = 2.0$ rad/s, $\zeta_0 = 0.05$ and $\mu = 200$ m represent the natural frequency, damping ratio, and nonlinearity coefficient, respectively. Finally, through comparison with the second-order statistical values obtained using the Monte Carlo Method (MCM) (Fig. 2.7) and the cumulative distribution functions (CDFs) at different time instances (Fig. 2.8), the efficiency and accuracy of GPDEM in dealing with stochastic processes and reliability of nonlinear structural responses under stochastic seismic excitations are verified.

2.6.3 Verification Based on Stochastic Dynamic and Probabilitistic Analysis of Multilayer Slopes

High-dimensional random parameter samples were generated using the GF-bias optimization-based point selection method. By combining this approach with the GPDEM, a probabilistic analysis of multi-layered soil-rock engineering slopes considering the randomness of material parameters was performed. By comparing the results with those obtained using traditional Monte Carlo simulation methods, this

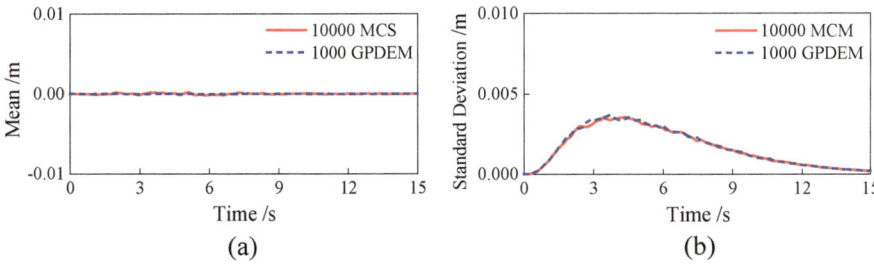

Fig. 2.7 Comparison of second-order statistical values between the GPDEM and MCM, **a** mean; **b** standard deviation

Fig. 2.8 Comparison of CDFs between the GPDEM and MCM

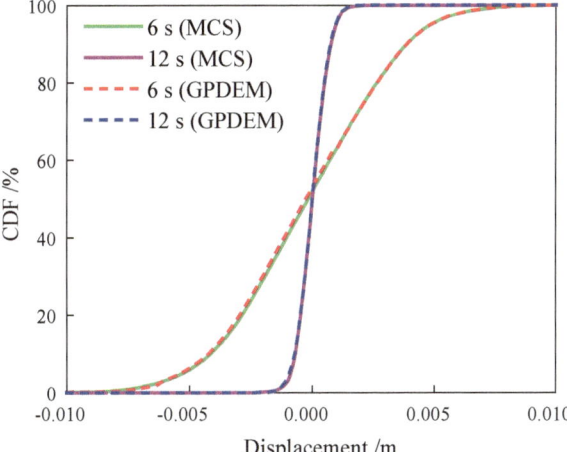

approach proves to be effective in generating samples and conducting structural reliability analysis involving dozens or even scores of random variables. It demonstrates high efficiency and reliability. The probabilistic analysis employed deterministic seismic motion, with acceleration time histories shown in Fig. 2.9, where the peak ground acceleration PGA = 0.2g.

Fig. 2.9 Acceleration time history of ground motion

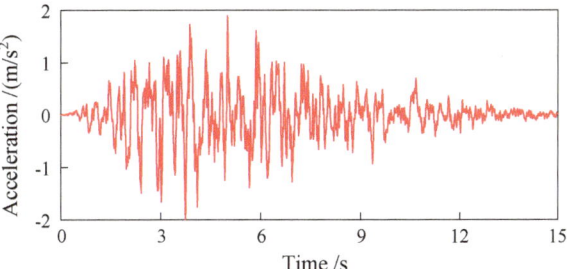

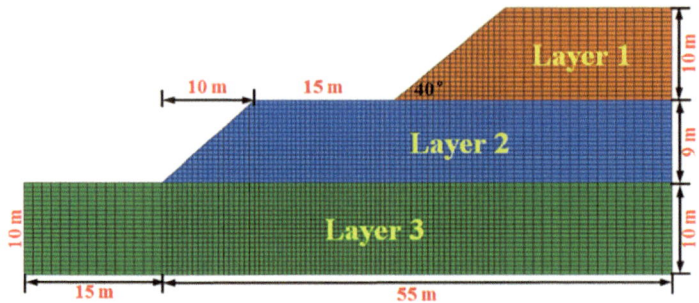

Fig. 2.10 Finite element mesh for case study 1

(1)   Case 1: Slope Model with Six Random Parameter

The first example involves a multi-layered embankment, as described in reference (Reale et al. 2016). Figure 2.10 shows its finite element mesh, consisting of two layers of fill material, topped with a 10 m-deep layer of hard silty soil. The parameter details are provided in Table 2.1. The mesh comprises 3985 elements and 4107 nodes. All random parameters are assumed to follow independent normal distributions, totaling six random variables. Monte Carlo simulation was performed 10,000 times, while GPDEM simulation was carried out 600 times.

     Figure 2.11 presents the stochastic process of safety factors, including a comparison of means and standard deviations. Figure 2.12 showcases probability density functions and cumulative distribution functions at several representative time instances. Figure 2.13 illustrates the probability density function (PDF) and cumulative distribution function (CDF) of the minimum safety factor. The reliability based on the Monte Carlo Method (MCM) and GPDEM is 0.8159 and 0.8173, respectively. It can be observed that the two results are in good agreement, yet the GPDEM demonstrates higher efficiency.

(2)   Case 2: Eight Random Parameter Slopes

The second example involves a complex four-layer soil profile, detailed in reference (Zolfaghari et al. 2005). Figure 2.14 depicts its finite element mesh. There are a total of eight random parameters, each following an independent lognormal distribution,

Table 2.1 Statistical values of soil parameters in case study 1

Layer	Bulk density/ (KN/m³)	Young's modulus/ MPa	Poisson's ratio	Cohesion/KPa		Angle of friction/°	
				Mean	Coefficient of Variation	Mean	Coefficient of Variation
Layer 1	18.0	20	0.3	10	0.1	28	0.05
Layer 2	18.5	20	0.3	8	0.15	29	0.1
Layer 3	20.0	20	0.3	5	0.2	36	0.1

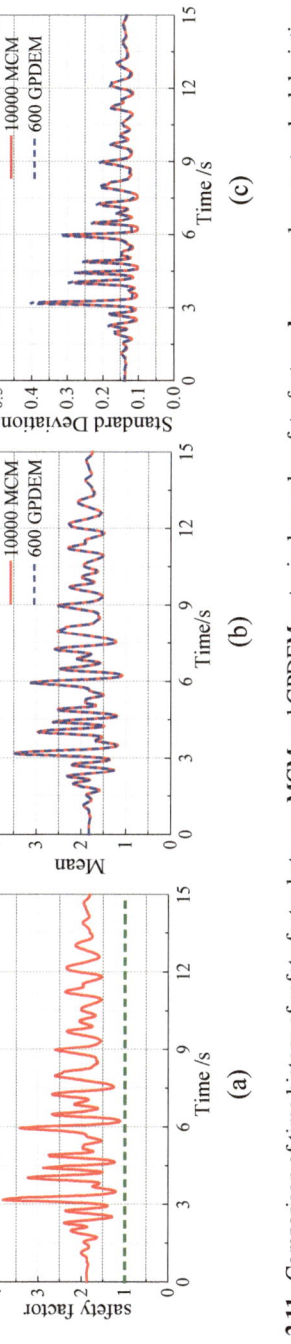

Fig. 2.11 Comparison of time history of safety factor between MCM and GPDEM, **a** typical sample safety factors; **b** mean values; **c** standard deviations

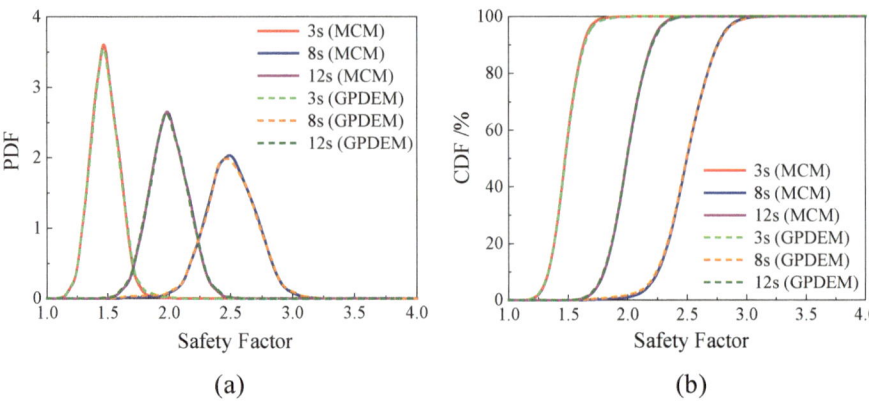

Fig. 2.12 Comparison of PDF and CDF of safety factor at typical instant between MCM and GPDEM, **a** PDF; **b** CDF

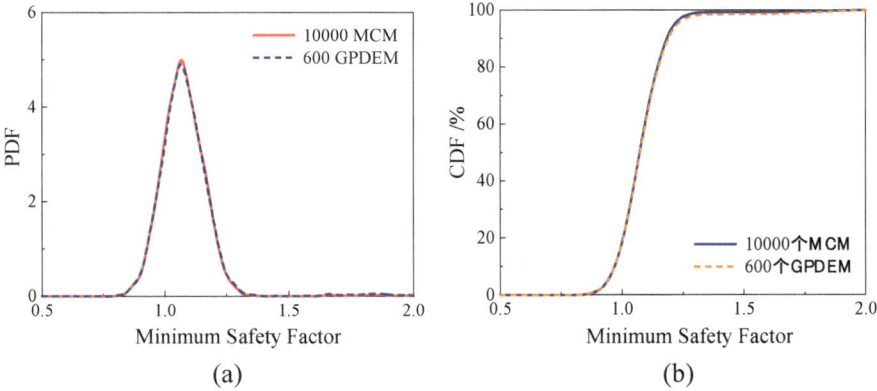

Fig. 2.13 Comparison of PDF and CDF of minimum safety factor between MCM and GPDEM, **a** PDF; **b** CDF

as specified in Table 2.2. Monte Carlo simulation was conducted 10,000 times, while GPDEM simulation was performed 800 times.

Figure 2.15 shows the stochastic process of safety factors, including a comparison of means and standard deviations. Figure 2.16 presents the probability density functions and cumulative distribution functions at several representative time instances. Figure 2.17 illustrates the probability density function and cumulative distribution function of the minimum safety factor. The reliabilities based on the MCM and GPDEM are 0.6701 and 0.6675, respectively.

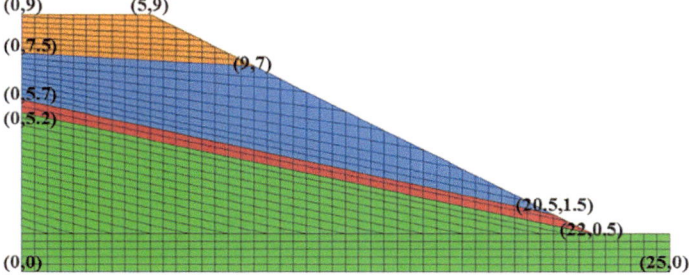

Fig. 2.14 Finite element mesh for case study 2

Table 2.2 Statistical values of soil parameters in case study 2

Layer	Bulk density/ (KN/m^3)	Young's modulus/ MPa	Poisson's ratio	Cohesion/KPa		Angle of friction/°	
				Mean	Coefficient of variation	Mean	Coefficient of variation
Layer 1	19.0	20	0.3	18	0.5	16	0.3
Layer 2	19.0	20	0.3	20	0.5	29	0.3
Layer 3	19.0	20	0.3	12	0.3	36	0.2
Layer 4	19.0	20	0.3	20	0.5	36	0.3

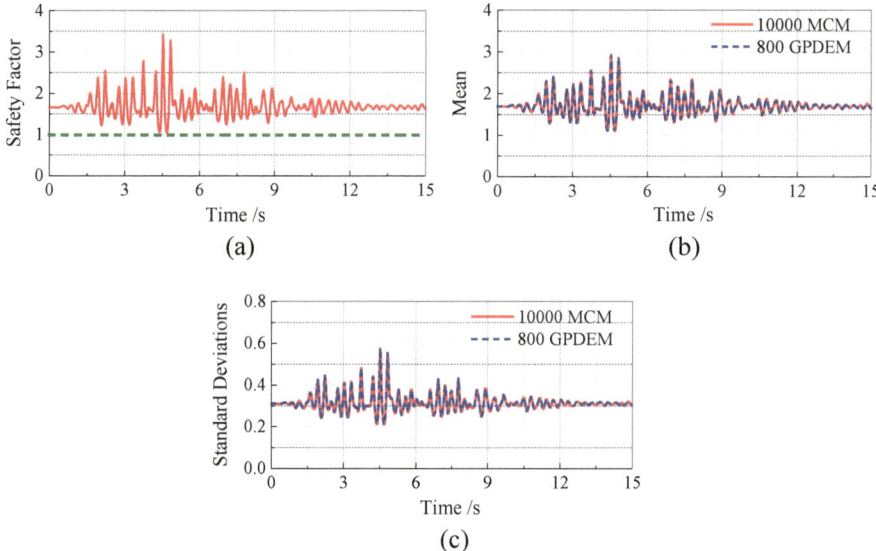

Fig. 2.15 Comparison of time history of safety factor between MCM and GPDEM, **a** typical sample safety factors; **b** mean values; **c** standard deviations

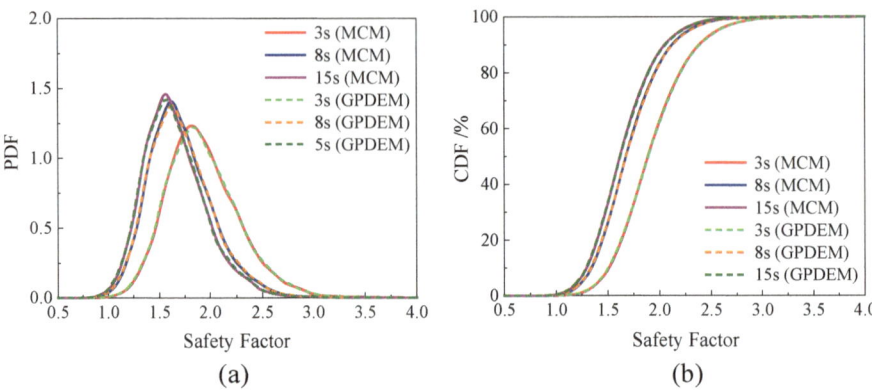

Fig. 2.16 Comparison of PDF and CDF of minimum safety factor between MCM and GPDEM, **a** PDF; **b** CDF

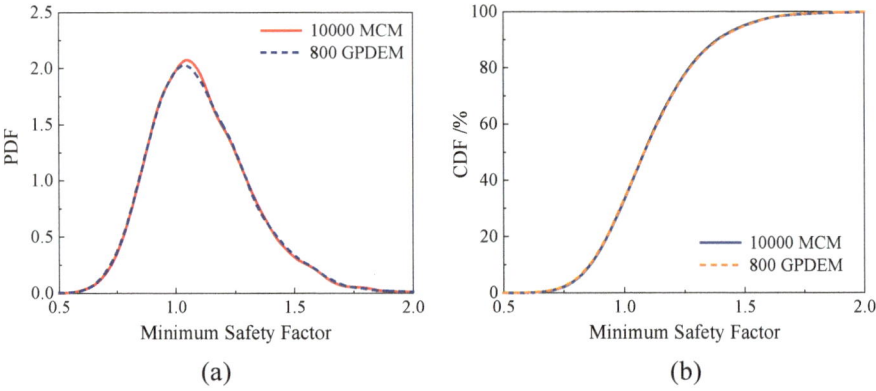

Fig. 2.17 Comparison of PDF and CDF of minimum safety factor between MCM and GPDEM, **a** PDF; **b** CDF

2.6.4   Verification Based on Stochastic Dynamic and Probabilistic Analysis of CFRD

To further verify the effectiveness and efficiency of the GPDEM combined with the non-stationary seismic generation method for random dynamic simulation and probabilistic analysis in highly nonlinear structural panel-stacked rockfill dams, this section contrasts the results with those obtained using traditional MCM. It examines their influence on the dam crest acceleration, panel stress stochastic dynamic response, and associated probabilities.

The finite element mesh of the concrete panel-stacked rockfill dam is shown in Fig. 2.18. In the calculations, the elements use quadrilateral 4-node isoparametric elements, with a total of 3777 elements and 3795 nodes. The computations were

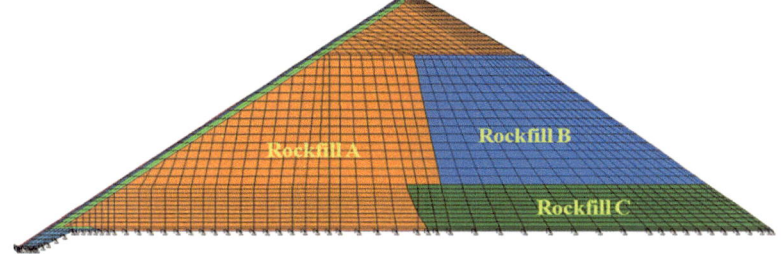

Fig. 2.18 Finite element mesh of the CFRD

conducted on a PC machine with an Intel i7 8-core CPU and 32GB of RAM. The dynamic calculation for a single sample takes about 3 min, while the stable calculation takes around 30 s. The dynamic water pressure is applied using the Westergaard additional mass method (Westergaard 1933). The considered panel-stacked rockfill dam has a crest width of 25 m, a height of 245 m, an upstream dam slope of 1:1.5, an upper downstream ramp slope of 1:1.7, and a lower downstream ramp slope of 1:1.4. The dam body consists of panels and five stacking areas: stacking area A, stacking area B, stacking area C, transition zone, and cushion zone. The reservoir water level is 225 m.

The static analysis of the rockfill material uses the Duncan E-B nonlinear elastic model (Duncan and Chang 1970), and Table 2.3 provides the parameters for the static model. The dynamic analysis employs the Hardin-Drnevich equivalent linear viscoelastic model (Hardin and Drnevich 1972), with parameters as presented in Table 2.4. The peak ground acceleration of the seismic motion is adjusted to PGA = 0.4g, and a total of 377 and 5000 seismic records are generated using the number theoretic method and the Monte Carlo method, denoted as 377 GPDEM and 5000 MCM, respectively. Through a series of finite element dynamic time history analyses and solutions of probability density evolution equations, the dam crest acceleration and panel stress stochastic dynamic response second-order statistical values, as well as probability information, can be obtained.

The comparison of mean and standard deviation time histories of dam crest acceleration and panel dynamic stress (tensile stress is positive) based on the Generalized Probability Density Evolution Method (GPDEM) and the Monte Carlo Method

Table 2.3 Parameters for Duncan E-B model

Material	$\rho/(kg/m^3)$	K	n	R_f	K_b	m	$\varphi_0/(°)$	$\Delta\varphi/(°)$
Rockfill A	2150	1109	0.24	0.64	420	0.26	49.8	7.2
Rockfill B	2100	800	0.32	0.64	490	0.30	49.8	7.2
Rockfill C	2170	980	0.26	0.79	400	0.31	50	8.2
Transition	2222	1250	0.31	0.78	500	0.16	53.5	10.7
Cushion	2258	1200	0.30	0.75	680	0.15	54.4	10.6

Table 2.4 Parameters for Hardin-Drnevich model

Material	K	n	v
Rockfill A	2660	0.444	0.33
Rockfill B	3115	0.396	0.33
Rockfill C	4997	0.298	0.33
Transition	3223	0.455	0.40
Cushion	3828	0.345	0.40

(MCM) is illustrated in Fig. 2.19. By comparing 377 GPDEM samples with 5000 MCM samples, the mean and standard deviation obtained from 377 GPDEM closely match those from 5000 MCM. This demonstrates the effectiveness and accuracy of the GPDEM method, which requires fewer finite element samples and computation time.

However, the fluctuations in the mean and standard deviation time histories also indicate significant differences in various physical quantities for different seismic responses. These differences arise from the stochastic nature of seismic motion. Therefore, a comprehensive analysis of the seismic response of high concrete faced rockfill dams is needed from the perspective of random vibration. Nevertheless, it's worth noting that there is still a slight difference between the results of 377 GPDEM samples and 5000 MCM samples, as seen from the second-order statistical values.

A series of deterministic seismic response analyses were used to obtain the physical quantities of acceleration and panel stress, which were then inserted into the

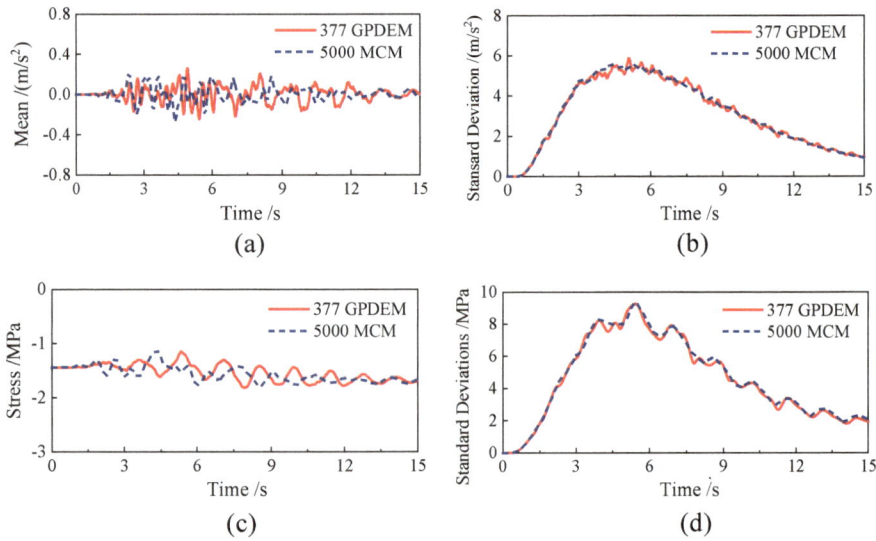

Fig. 2.19 Comparison of mean and standard deviation of acceleration and stress between GPDEM and MCM, **a** mean acceleration; **b** acceleration standard deviation; **c** time history of stress mean; **d** time history of stress standard deviation

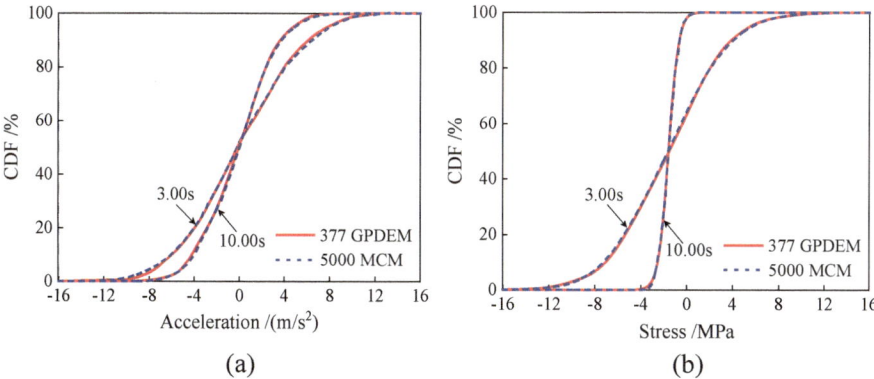

Fig. 2.20 Comparison of dynamic CDFs of dynamic acceleration and stress at typical instant between GPDEM and MCM, **a** CDF of maximum acceleration; **b** CDF of maximum stress

GPDEM equation. The probability information was obtained using the Total Variation Diminishing (TVD) scheme of finite difference method. Figure 2.20 shows the cumulative distribution functions at two representative time instances. It can be observed that the cumulative distribution functions obtained from 377 GPDEM samples and 5000 MCM samples fit well, confirming the high precision and effectiveness of GPDEM. It's also noticeable that the cumulative distribution functions change over time, indicating the significant influence of seismic motion on the seismic response of panel-stacked rockfill dams under the coupling effect of nonlinear rockfill behavior and stochastic seismic excitation. Therefore, the seismic response of high concrete faced rockfill dams requires analysis from a stochastic perspective.

Figure 2.21 presents the cumulative distribution functions of maximum acceleration and maximum stress obtained based on the virtual random process and solving the GDEE using the SUPG scheme of the finite element method. The comparison between the results of 377 GPDEM samples and 5000 MCM samples further validates the accuracy and efficiency of the GPDEM method. Additionally, due to the influence of the probability density evolution process, an improved level of accuracy is achieved.

By comparing the failure probabilities or reliabilities, as well as the mean and standard deviation of responses, the results obtained from several hundred GPDEM simulations are within the same order of magnitude as those obtained from tens of thousands of Monte Carlo simulations. However, the efficiency of GPDEM is dozens or even several tens of times higher than the Monte Carlo method. This indicates the effectiveness and efficiency of this approach for large-scale geotechnical engineering stochastic and probabilistic analysis. In conclusion, considering the high accuracy and efficiency of the Generalized Probability Density Evolution Method, along with its strong theoretical foundation, it can be effectively employed for performance-based seismic safety assessment of high concrete faced rockfill dams, with the expectation of achieving favorable results.

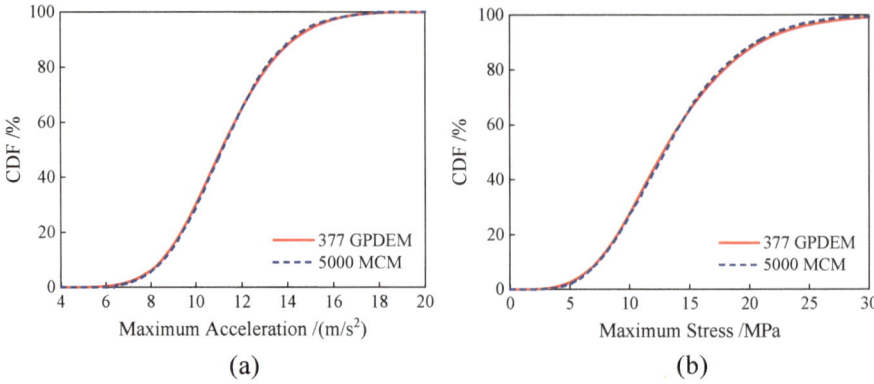

Fig. 2.21 Comparison of CDFs of maximum acceleration and stress between GPDEM and MCM, **a** CDF of maximum acceleration; **b** CDF of maximum stress

2.7　Seismic Fragility Analysis

Seismic vulnerability analysis is a crucial step in the next-generation performance-based seismic design and a natural extension of the performance-based design philosophy. Seismic vulnerability typically refers to the probability of a structure reaching a certain state of damage under a known seismic intensity. It quantitatively expresses the structural seismic performance from a probabilistic perspective, reflecting the probabilistic relationship between seismic intensity and the extent of structural damage.

There are four main methods for conducting structural seismic vulnerability analysis: judgment methods, empirical methods, experimental methods, and numerical analysis methods. Judgment methods involve a broad assessment and judgment of different types of structural damage within a region based on the expertise of experts and engineers. The concept was introduced by American scholar Whiteman in 1973, who proposed the Damage Probability Matrix (DPM) method. Empirical methods use a large amount of observed structural damage data to predict the probability of various levels of damage occurring under different seismic intensity levels, resulting in empirical vulnerability curves.

Experimental methods establish physical structural models and study vulnerability through extensive experimental testing. However, these methods are often limited by sample quantity, laboratory and equipment conditions. Considering the seismic conditions of dams, especially high concrete faced rockfill dams, the first three methods are generally less applicable. Damages to high concrete faced rockfill dams are rare, and the complexity of dam construction materials, loading conditions, and boundary conditions makes it difficult to simulate realistic damage through laboratory experiments.

Therefore, numerical analysis methods should become an effective approach for vulnerability analysis, including that of high concrete faced rockfill dams. Numerical analysis is widely used in the field of structural seismic performance research. It involves constructing numerical analysis models using finite element methods, selecting actual or artificially generated seismic motions, and conducting numerous numerical simulations to obtain the seismic response of structures. This information is then used to derive vulnerability analysis curves. While this method is widely used in large civil engineering projects such as buildings, bridges, concrete dams, and nuclear power structures, its application to earth-rock dams, especially high concrete faced rockfill dams, has been limited. The detailed analysis process can be represented by Fig. 2.22.

The results of seismic vulnerability analysis for structures are typically presented in two ways: vulnerability curves and vulnerability matrices. Figure 2.23 illustrates a typical form of vulnerability curve for structures, and Table 2.5 presents a typical vulnerability matrix obtained by Liang et al. through the analysis of dam damage data from the Wenchuan earthquake in Mianyang City.

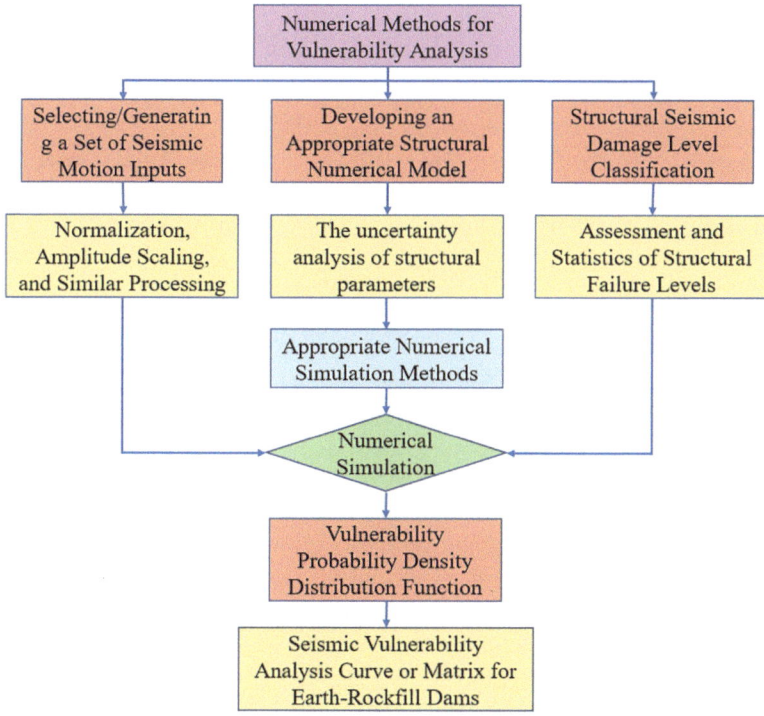

Fig. 2.22 Flow chart of numerical method for fragility analysis

Fig. 2.23 Typical structural fragility curve

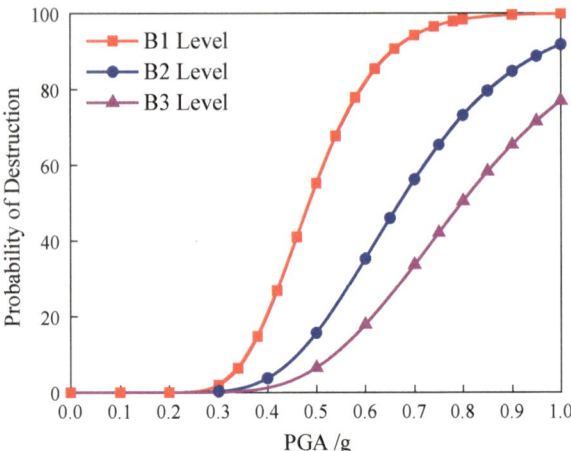

Table 2.5 Damage probability matrix for earth-rock dams in Mianyang

Seismic damage condition	Loss ratio range/%	Median loss ratio/%	The probability of dam failure at different intensity levels/%			
			VI	VII	VIII	IX
Intact	0–10	5	49.77	29.17	1.75	0
Moderate risk situation	10–20	15	39.73	44.55	62.81	33.33
High risk situation	20–50	35	10.05	22.76	27.02	44.44
Dam breach hazard	50–70	60	0.46	3.53	8.42	22.22
Earthquake destruction	70–100	85	0	0	0	0

2.8 Conclusion

This chapter provides an overview of uncertainty factors and some probabilistic analysis methods in seismic response analysis of earth-rock dams. These include traditional methods like the first and second-order moment method, Monte Carlo method, and response surface method. The chapter also covers the recently popular research topics, such as the probability density evolution method based on stochastic vibration theory and non-stationary stochastic seismic motion models. Additionally, it introduces the vulnerability analysis process suitable for performance-based seismic safety assessment of high concrete faced rockfill dams.

From the above research, it's evident that there is limited work related to probabilistic analysis of earth-rock dams, particularly using stochastic dynamic time history analysis methods, and rare studies focused on high concrete faced rockfill dams. The uncertainty factors in earth-rock dams often consider stochastic seismic excitations and often overlook the influence of dam construction materials, which should not be neglected. Few studies fully consider the stochastic process and dynamic probability of seismic response in earth-rock dams.

Furthermore, this chapter elaborately explains the Generalized Probability Density Evolution Method and its application and solving process, as well as the process of generating stochastic seismic motion and high-dimensional random sample parameters. It establishes a non-stationary stochastic seismic motion model based on the latest seismic design code spectrum, and discretely generates seismic acceleration sample time histories with rich probabilistic characteristics. Through various equivalent linear stochastic dynamic and probabilistic analyses including analytical solutions, Duffing oscillator, multi-layered soil slopes, and panel-stacked rockfill dams, the combination of stochastic seismic motion and high-dimensional random parameter generation methods with the Generalized Probability Density Evolution Method is validated. This approach not only exhibits high efficiency but also guarantees a high level of accuracy in analyzing the random dynamic response and probabilistic behavior of complex geological and engineering structures. This paves the way for subsequent research in the field of elastoplastic stochastic dynamic analysis and probabilistic assessment for high concrete faced rockfill dams, as well as for performance-based seismic safety evaluations, laying a solid foundation.

References

Box GEP, Wilson KB (1951) On the experimental attainment of optimum conditions. J Roy Stat Soc: Ser B (methodol) 13:1–38

Bucher CG, Bourgund U (1990) A fast and efficient response surface approach for structural reliability problems. Struct Saf 7:57–66

Cacciola P, Deodatis G (2011) A method for generating fully non-stationary and spectrum-compatible ground motion vector processes. Soil Dyn Earthq Eng 31:351–360

Chen JB, Li J (2014) Probability density evolution method for stochastic seismic response and reliability of structures. Eng Mechan 31:1–10

Chen GX, Xie JF, Zhang KX (1995) One-dimensional random seismic response analysis method for heterogeneous earth dams. Eng Reliab 35–38

Chen JB, Li J (2009) A note on the principle of preservation of probability and probability density evolution equation. Probab Eng Mech 24:51–59

Chen JB, Yang JY, Li J (2016) A GF-discrepancy for point selection in stochastic seismic response analysis of structures with uncertain parameters. Struct Saf 59:20–31

Cho SE (2009) Probabilistic stability analyses of slopes using the ANN-based response surface. Comput Geotech 36:787–797

Clough RW (1993) Dynamics of structures, 2nd edn. McGraw-Hill, New York

Das PK, Zheng Y (2000) Cumulative formation of response surface and its use in reliability analysis. Probab Eng Mech 15:309–315

Deng J, Gu DS, Li XB (2005) Structural reliability analysis for implicit performance functions using artificial neural network. Struct Saf 27:25–48

Du XL, Chen HQ (1994) Random simulation and its parameter determination method of earthquake ground motion. Earthq Eng Eng Vib 1–5

Duncan JM, Chang CY (1970) Nonlinear Anal Stress Strain Soils. J Soil Mechan Found Div 96:1629–1653

El Hami A, Radi B (2016) Stochastic dynamics of structures. Wiley

Elman H, Silvester D, Wathen A (2014) Finite elements and fast iterative solvers: with applications in incompressible fluid dynamics, 2nd edn. Oxford University Press

Fujimura K, Der Kiureghian A (2007) Tail-equivalent linearization method for nonlinear random vibration. Probab Eng Mech 22:63–76

Haian L (2013) Research on seismic damage prediction and rapid evaluation method of earth-rock dam. Int J Earthq Dyn 44–45

Hardin BO, Drnevich VP (1972) Shear modulus and damping in soils: design equations and curves. J Soil Mechan Found Div 98:667–692

Hasofer A, Lind N (1974) Exact and invariant second-moment code format. J Eng Mechan Div-ASCE 100:111–121

Housner GW (1947) Characteristics of strong-motion earthquakes. Bull Seismol Soc Am 37

Hughes TJR, Franca LP, Balestra M (1986) A new finite element formulation for computational fluid dynamics: V. circumventing the Babuška-Brezzi condition: a stable Petrov-Galerkin formulation of the Stokes problem accommodating equal-order interpolations. Comput Method Appl Mechan Eng 59:85–99

Kanai K (1957) Semi-empirical formula for the seismic characteristics of ground. Trans Archit Inst Jpn 57(1):281–284

Kartal ME, Bayraktar A, Başağa HB (2010) Seismic failure probability of concrete slab on CFR dams with welded and friction contacts by response surface method. Soil Dyn Earthq Eng 30:1383–1399

Kaymaz I, McMahon CA (2005) A response surface method based on weighted regression for structural reliability analysis. Probab Eng Mech 20:11–17

Keng HL, Yuan W (1981) Applications of number theory to numerical analysis. Heidelberg, Berlin

Li J, Chen JB (2003) Probability density evolution method for analysis of stochastic structural dynamic response. Theor Appl Mechan 437–442

Li J, Chen JB (2010) Advances in the research on probability density evolution equations of stochastic dynamical systems. Adv Mechan 40:170–188

Li J, Chen JB (2017) Some new advances in the probability density evolution method. Appl Math Mechan 38:32–43+2

Lin JH, Zhong WX (1998) Some notes on FEM and structural random response analysis. Chin J Comput Mechan 94–100

Liu HL (1996) Permanent deformation of foundation and embankment dam due to stochastic seismic excitation. Chin J Geotech Eng 19–27

Liu LB (2013) Reliability analysis of nonlinear stochastic structures based on probability density evolution (Master's thesis). Dalian University of Technology

Liu ZJ, Fang X (2012) Stochastic earthquake response and seismic reliability analysis of large-scale aqueduct structures. J Yangtze River Sci Res Inst 29:77–81

Liu HL, Lu ZQ, Qian JH (1996) Nonlinear random response and dynamic reliability analysis of earth-rock dam. J Hohai Univ 105–109

Liu ZJ, Lei YL, Fang X (2013) Probability density evolution method-based random seismic response analysis of cable-stayed bridges. Chin Civil Eng J 46:226–232

Liu ZJ, Zeng B, Zhou YH et al (2014) Probabilistic model of ground motion processes and seismic dynamic reliability analysis of the gravity dam. J Hydraul Eng 45:1066–1074

Luo XF, Li X, Zhou J et al (2012) A Kriging-based hybrid optimization algorithm for slope reliability analysis. Struct Saf 34:401–406

Ou JP, Niu DT (1990) Parameters in the random process models of earthquake ground motion and their effects on the response of structures. J Harb Univ Civil Eng Archit 24–34

Ou J, Wang G (1998) Random vibration of structures. Higher Education Press, Beijing

Radović I, Sobol' IM, Tichy RF (1996) Quasi-Monte Carlo methods for numerical integration: comparison of different low discrepancy sequences. Monte Carlo Methods Appl 2:1–14

Reale C, Xue J, Gavin K (2016) System reliability of slopes using multimodal optimization. Géotechnique 66:413–423

Sanchez Lizarraga H, Lai CG (2014) Effects of spatial variability of soil properties on the seismic response of an embankment dam. Soil Dyn Earthq Eng 64:113–128

Sekar P, Narayanan S (1994) Periodic and chaotic motions of a square prism in cross-flow. J Sound Vib 170:1–24

Shao LT, Tang HX, Kong XJ et al (1999) Finite element analysis for slope stability of earth-rock dam under the action of stochastic seismic. J Hydraul Eng 66–71

Wang DB, Liu HL, Yu T et al (2013) Seismic fragility analysis for earth-rock dams based on deformation. Chin J Geotech Eng 35:814–819

Wang ZH, Liu HL, Chen GX (2006) Study on stationary stochastic seismic motion model and Stochastic Earthquake response of earth-rock dam. J Disas Preven Mitig Eng 389–394

Westergaard HM (1933) Water pressures on dams during earthquakes. Trans Am Soc Civ Eng 98:418–433

Wichtmann T, Triantafyllidis T (2013) Effect of uniformity coefficient on G/G_{max} and damping ratio of uniform to well-graded quartz sands. J Geotech Geoenviron Eng 139:59–72

Wong FS (1985) Slope reliability and response surface method. J Geotech Eng 111:32–53

Wu ZG (1991) Seismic response of soil layers with random material parameters. J Hydraul Eng 1–6

Yang G, Zhu S (2016) Seismic response of rockfill dams considering spatial variability of rockfill materials via random finite element method. Chin J Geotech Eng 38:1822–1832

Yu SD, Du JH, Xiong XR (1993) The stochastic dynamic reliability analysis of earth dam. J Yangtze River Sci Res Inst 68–72

Zhang GW, Liu LY (1994) Basic parameter probability and correlation analysis for earth-rock dam materials. Water Resour Hydropower Eng 12–16

Zhao HB (2008) Slope reliability analysis using a support vector machine. Comput Geotech 35:459–467

Zhu WQ (1993) Stationary solutions of stochastically excited dissipative Hamiltonian systems. Chin J Theor Appl Mechan 676–684

Zolfaghari AR, Heath AC, McCombie PF (2005) Simple genetic algorithm search for critical non-circular failure surface in slope stability analysis. Comput Geotech 32:139–152

Chapter 3
Stochastic Dynamic Analysis of CFRD Considering Randomness of Ground Motion

3.1 Introduction

With the improvement in the efficiency and accuracy of finite element numerical calculations, the nonlinear time history analysis method has gradually become the mainstream seismic safety assessment approach in the field of earth-rock dam engineering. However, seismic loads exhibit significant uncertainties, leading to a certain level of randomness in the seismic response of high CFRD. The results obtained from seismic time history analysis vary significantly with different ground motions or intensity levels. Therefore, relying on only a few seismic records makes it challenging to comprehensively understand the seismic performance of high CFRD. On the other hand, as the theory of performance-based seismic safety assessment continues to evolve, there is a need to understand the performance levels and seismic safety of complex and critically important engineering structures like high CFRD under different seismic intensities. Traditional deterministic analysis methods are insufficient for meeting such requirements. Thus, there is a need to delve into the performance levels of dams under future seismic actions from the perspective of fragility. The purpose of probability analysis is to predict the probability of high CFRD reaching various performance levels under different ground motions or intensity levels. It is a crucial component of performance-based seismic safety assessment, allowing for the consideration of uncertainties in structural seismic responses. Currently, there is limited research on the seismic safety assessment of rockfill dams, especially high CFRD, from a probabilistic perspective.

However, it is well known that many 200 m or even 300 m high CFRDs are under construction or planned in the western regions of China, and they are located in high-intensity seismic zones. Therefore, it is necessary to gradually improve their seismic safety assessment system and establish a performance-based seismic safety assessment method. However, the dynamic response and failure modes of high CFRDs under seismic action are complex. There are few reported instances of seismic damage to high CFRDs, especially from different seismic intensities,

B. Xu and R. Pang, *Stochastic Dynamic Response Analysis and Performance-Based Seismic Safety Evaluation for High Concrete Faced Rockfill Dams*, Hydroscience and Engineering, https://doi.org/10.1007/978-981-97-7198-1_3

making the study of seismic damage modes and their quantitative description, analysis, and classification still in the preliminary research stage. Furthermore, the rockfill material of high CFRDs exhibits strong nonlinear characteristics, and seismic, especially strong seismic, actions have a significant impact on its response. Therefore, advanced numerical models need to be selected to simulate its dynamic response. Lastly, because probabilistic analysis requires many finite element dynamic time history analysis considering various uncertainties, it is challenging to analyze large-volume, strongly nonlinear high CFRDs or the analysis takes a long time. Therefore, precise and rapid probabilistic analysis methods need to be adopted.

In this chapter, based on the above problems, the stochastic ground motion is fully considered, and the spectral expression-random function method is used to obtain the sample timescales of ground motion acceleration with rich probabilistic features in the same set system and the assigned probability corresponding to each ground motion. Through the high-performance geotechnical engineering nonlinear dynamic analysis program, GEODYNA, developed by the research team, a series of finite element elastoplastic dynamic analyses are conducted on high CFRDs. Combined with the generalized probability density evolution method, it obtains information on random dynamic responses and probability distributions. This forms the preliminary framework for a performance-based seismic safety assessment under random seismic actions. Firstly, it examines the random dynamic and probabilistic response patterns of several commonly used response indicators for high CFRDs based on elasto-plastic analysis. This includes dam body acceleration, deformation, and panel stress. From the perspectives of random dynamics and probability, numerical distribution ranges for these response indicators under different seismic intensities are suggested, providing reference values for seismic design and ultimate seismic capacity analysis of high CFRDs. Subsequently, based on performance indicators, such as dam top settlement deformation and the panel demand-to-capacity ratio, combined with the super-stress duration, preliminary recommendations for corresponding performance levels are provided. The study establishes a seismic safety assessment method based on a multi-seismic intensity, multi-performance target, and exceedance probability performance relationship, along with fragility probability curves.

3.2 Computational Constitutive Models

This chapter primarily applies an improved generalized plasticity constitutive model for rockfill, based on the generalized plasticity theory, boundary surface theory, and critical state theory. This model can better reflect the actual conditions of soil, including cyclic hardening–softening, shear dilation-shrinkage, and particle crushing. It can analyze the entire process of seismic dynamic response and permanent deformation, making it theoretically more reasonable (Feng et al. 2010). On the other hand, to address the contact issues between the rockfill and the panel under dynamic loading, some contact constitutive models corresponding to the generalized plasticity model have also been proposed and developed, such as the generalized plasticity contact surface model (Liu et al. 2012).

3.2.1 Generalized Plastic Static and Dynamic Unified Model for Rockfill

To address the requirements of significant variations in the average principal stress, Dalian University of Technology has implemented a series of enhancements to the original generalized plasticity model (Xu et al. 2012; Zou et al. 2013; Liu et al. 2015) These improvements take into account the stress correlation between the elastic modulus and loading/unloading modulus of dam construction materials, enabling effective consideration of phenomena such as shear expansion, shear contraction, and cyclic cumulative deformation under dynamic conditions. The framework is clear and utilizes a set of parameters to facilitate static and dynamic simulation and analysis of rock pile materials.

The generalized elastoplastic stress increment of the generalized plasticity model can be expressed as:

$$d\boldsymbol{\sigma} = \mathbf{D}^{\mathrm{ep}}:d\boldsymbol{\varepsilon} \tag{3.1}$$

The strain increment consists of both elastic strain increment $d\varepsilon_e$ and plastic strain increment $d\varepsilon_p$:

$$d\boldsymbol{\varepsilon}_e = \boldsymbol{C}^e:d\boldsymbol{\sigma}$$

$$d\boldsymbol{\varepsilon}_p = \frac{1}{H_{\mathrm{L/U}}}\mathbf{n}_{\mathrm{gL/U}} \otimes \mathbf{n}:d\boldsymbol{\sigma}$$

The matrix expression can be represented as:

$$\mathbf{D}^{\mathrm{ep}} = \mathbf{D}^e - \frac{\mathbf{D}^e : \mathbf{n}_{\mathrm{gL/U}} \otimes \mathbf{n} : \mathbf{D}^e}{H_{\mathrm{L/U}} + \mathbf{n} : \mathbf{D}^e : \mathbf{n}_{\mathrm{gL/U}}} \tag{3.2}$$

where $\mathbf{D}^{\mathrm{ep}}$ represents the elastoplastic matrix, which is influenced by factors such as the current stress state, stress level, stress history, loading and unloading directions, and changes in the microstructure of particles. L and U represent loading and unloading, respectively. $\mathbf{D}^e$ represents the elastic matrix. $\mathbf{n}_{\mathrm{gL}}$ and $\mathbf{n}_{\mathrm{gU}}$ represent the plastic flow directions during loading and unloading, respectively, and they signify the direction of plastic strain increments. $\mathbf{n}$ is the loading direction vector, representing the direction of the yield surface normal. H_{L} and H_{U} are the plastic moduli during loading and unloading, respectively.

The generalized plasticity P-Z model was primarily proposed to address the liquefaction problem in sandy soils. In the analysis of soil liquefaction, the variation in confining pressure is relatively small. However, with high earth-rock dams, there is a significant variation in the average principal stress within the dam body. When considering pressure dependency in the P-Z model, its parameters are greatly influenced by the average principal stress. Therefore, there are some limitations in the application of this model for static and dynamic analysis of high earth and rock dam. Dalian University of Technology has made improvements to the P-Z model, and the modified values of the plastic modulus includes loading and unloading modulus and elastic modulus are as follows:

$$H_{\mathrm{L}} = H_0 \cdot p_{\mathrm{a}} \cdot \left(p/p_{\mathrm{a}}\right)^{m_1} \cdot H_{\mathrm{f}} \cdot (H_{\mathrm{v}} + H_{\mathrm{s}}) \cdot H_{\mathrm{DM}} \cdot H_{\mathrm{den}} \tag{3.3}$$

$$H_{\mathrm{u}} = \begin{cases} H_{\mathrm{u0}} p_{\mathrm{a}} \left(p/p_{\mathrm{a}}\right)^{m_{\mathrm{u}}} \left(\eta_{\mathrm{u}}/M_{\mathrm{g}}\right)^{-\gamma_{\mathrm{u}}} & \left|\eta_{\mathrm{u}}/M_{\mathrm{g}}\right| < 1 \\ H_{\mathrm{u0}} & \left|\eta_{\mathrm{u}}/M_{\mathrm{g}}\right| \ge 1 \end{cases} \tag{3.4}$$

$$K = K_0 p_{\mathrm{a}} (p/p_{\mathrm{a}})^{m_{\mathrm{v}}} \tag{3.5}$$

$$G = G_0 p_{\mathrm{a}} (p/p_{\mathrm{a}})^{m_{\mathrm{s}}} \tag{3.6}$$

In order to be able to better consider the hysteresis properties of the rock pile, the stress history function H_{DM} is also modified for reloading as:

$$H_{\mathrm{DM}} = \exp((1 - \eta/\eta_{\max})\gamma_{\mathrm{DM}}) \tag{3.7}$$

Meanwhile the cyclic densification effect has been considered, where $H_{\mathrm{den}} = e^{\gamma_{\mathrm{d}}\varepsilon_{\mathrm{v}}}$ represents the densification coefficient to consider the cyclic hardening characteristics of rockfill. Determine each parameter of the above model by particle swarm optimization algorithm, mainly including: G_0, m_{s}, K_0, m_{v}, α_{g}, α_{f}, M_{g}, M_{f}, H_0, m_1, β_0, β_1, H_{U0}, m_{u}, γ_{d}, γ_{DM}, γ_{U} seventeen parameters.

3.2.2 Generalized Plastic Interface Model

Liu et al. (2008) proposed a two-dimensional contact surface model based on the critical state and the generalized plasticity framework. Based on this model and the boundary surface theory, academician Kong et al. (2014) developed a static and dynamic unified two- and three-dimensional generalized plastic contact surface model for panel dams on the base of the above-mentioned generalized plasticity model of rockfill, which can effectively reflect the characteristics of the contact surface, such as shear expansion and contraction, hardening and softening, residual deformation, and particle crushing. As shown in Fig. 3.1, two boundary surfaces are defined on the normalized shear surface $\frac{\tau_x}{\sigma_n} - \frac{\tau_y}{\sigma_n}$, which is approximated as a

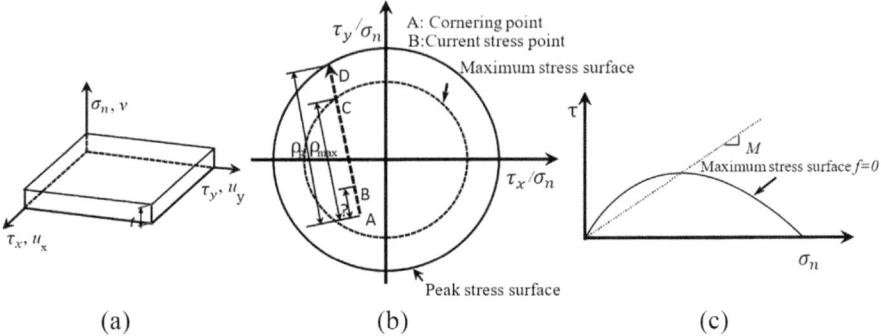

Fig. 3.1 Definitions of some basic concepts on the boundary surface, **a** stress–strain, **b** definition of boundary surface, **c** maximum stress surface in τ-σ_n space

circle. These surfaces include the peak stress boundary surface and the maximum stress history boundary surface. Additionally, the maximum stress history boundary surface is also defined in $\tau - \sigma_n$ space:

$$f = \tau - M\sigma_n \left(\frac{\alpha}{\alpha - 1}\right)\left[1 - \left(\frac{\sigma_n}{\sigma_c}\right)^{\alpha-1}\right] = 0 \qquad (3.8)$$

then, the relationship between the stress increment and displacement increment at the contact surface under three-dimensional conditions can be obtained and expressed as:

$$\begin{pmatrix} d\tau_x \\ d\tau_y \\ d\sigma_n \end{pmatrix} = \frac{1}{t}\mathbf{D}^{ep} \begin{pmatrix} du_x \\ du_y \\ dv \end{pmatrix} \qquad (3.9)$$

where @@$d\sigma = (d\tau_x, d\tau_y, d\sigma_n)^T$ is the stress increment, $d\sigma = (du_x, du_y, dv_n)^T/t$ is the strain increment, t is the thickness of the contact surface, generally equal to 5–10 times the average particle size.

The elastoplastic matrix is expressed as

$$\mathbf{D}^{ep} = \mathbf{D}^e - \frac{\mathbf{D}^e : \mathbf{n}_g \otimes \mathbf{n} : \mathbf{D}^e}{H + \mathbf{n} : \mathbf{D}^e : \mathbf{n}_g} \qquad (3.10)$$

Modification of the model loading and unloading judgment method is consistent with the generalized plasticity models:

$\mathbf{n} : d\sigma^e > 0$ represents loading;

$\mathbf{n} : d\sigma^e < 0$ represents unloading;

$\mathbf{n} : d\sigma^e = 0$ represents the neutral variable load, where $d\sigma^e = \mathbf{D}^e d\varepsilon$.

However, unlike the conventional generalized plasticity model, the stress state in the loading direction n is determined when there is a backbend point by replacing the absolute stress state $\bar{\sigma}$ by the stress state σ at the mapped point on the maximum stress boundary plane.

The model parameters include the elasticity parameter D_{n0}, D_{s0}; critical state parameter e_τ, λ, M_c; Plastic flow direction α, γ_d, k_m; Load Direction Parameters M_f; Plastic modulus parameters H_0, k, f_h; Particle Breaking Parameters a, b, c, $c_0 = 0.0001$, Most of these parameters can be determined directly from test results.

3.3 Ground Motion Input Method

With the increasing height of the dam, the limitations of the traditional consistency input method, which cannot consider the radiative damping effect of infinite foundations and the influence of traveling wave effects, become more and more prominent. Therefore, scholars at home and abroad have carried out in-depth research on artificial boundaries (Lysmer et al. 1969; Deeks et al. 1994; Liu et al. 1998, 2005, 2006) among which viscoelastic boundaries have been widely used in simulating dam-foundation interactions due to their ability to simultaneously simulate the scattering effect of waves and the elastic recovery of semi-infinite foundations and their ability to overcome the low-frequency drift caused by viscous boundaries. In this chapter, ground motion inputs are implemented using viscoelastic artificial boundaries and equivalent nodal loads, as shown in Fig. 3.2.

The boundary cell spring and damping coefficients are:

$$k_n = \alpha_n \frac{G}{r} \tag{3.11}$$

$$C_n = \rho v_p \tag{3.12}$$

$$k_t = \alpha_t \frac{G}{r} \tag{3.13}$$

$$c_t = \rho v_s \tag{3.14}$$

Fig. 3.2 Viscoelastic artificial boundary element

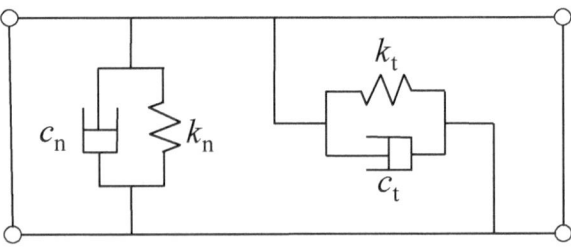

where v_p is the P-wave speed, $v_p = \sqrt{2(1-\mu)(1-2\mu)v_s}$, v_s is S-wave speed, $v_s = \sqrt{G/\rho}$.

The equivalent nodal load calculation expression is:

$$F_b = R_b^{ef} + C_b \dot{u}_b^{ef} + K_b u_b^{ef} \tag{3.15}$$

where $\dot{u}_b^{ef}, u_b^{ef}, R_b^{ef}$ are the velocity, displacement and force vectors induced by the free wave field at the boundary nodes, respectively; K_b, C_b is the additional stiffness matrix and the additional damping matrix, respectively; and F_b is the equivalent nodal load applied on the boundary.

3.4 Stochastic Dynamic Response and Probabilistic Analysis for High CFRDs

In this section, considering the randomness of ground motion, joint random ground motion generation method, generalized probability density evolution method, reliability probabilistic analysis method, susceptibility analysis method, and generalized plasticity model of rock pile material and contact surface, etc., stochastic dynamic and probabilistic analyses are carried out on a 250 m panel rockfill dam, to reveal the stochastic dynamic response law and damage probability under the action of different ground motion and different seismic intensities, to provide references for the seismic safety evaluation of CFRD based on their performance.

3.4.1 Finite Element Model and Material Parameter Information

The finite element grid of the concrete panel rockfill dam is shown in Fig. 3.3, with a dam height of 250 m, upstream dam slope gradient of 1:1.4, and downstream dam slope gradient of 1:1.6. The width of the top of the dam is 20 m, with a bedding zone and a transition zone below the panels, in which the width of the bedding zone is 3 m, the width of the transition zone is 4 m. According to "Code for Design of Concrete Face Rockfill Dams" (SL228-2013), the panel thickness is obtained as 0.30 + 0.0035 H) m, and H is the dam height. The panels are poured in three phases, 75, 150, and 250 m respectively. The dam body is filled in 50 layers, and the water is stored in 48 steps to 240 m. The unit is simulated by a quadrilateral unit, and the contact surface between the panel and the cushion layer is a Goodman element without thickness. The thickness and width of the dam foundation shall be 1/2 of the length of the dam bottom. After filling and impounding, perform dynamic calculations under seismic loads, in which static calculations provide initial stress, strain and displacement fields for dynamic calculations. The seismic input adopts

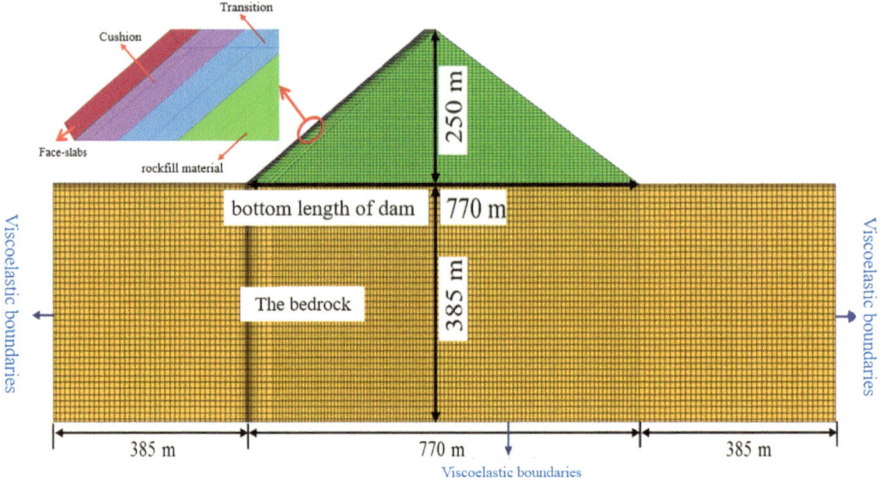

Fig. 3.3 Finite element mesh of the CFRD

the wave input method based on viscoelastic artificial boundary setting, and the hydrodynamic pressure on the panel is simulated by the additional mass method.

The generalized plastic model of rockfill materials is used to simulate the static and dynamic processes for rockfill materials, cushion materials, and transition materials, and the parameters are the numerical values from literature (Xu et al. 2012). Refer to Table 3.1. In the static and dynamic analysis, the contact surface between the panel and the cushion is simulated by the generalized plastic contact surface model, and the parameters are the values in the literature (Liu et al. 2014), refer to Table 3.2. Both are the dynamic parameters of the Zipingpu CFRD. The bedrock is simulated by a linear elastic model, and the parameters are density $\rho = 2600$ kg/m^3, elastic modulus $E = 2.0 \times 10^4$ MPa, and Poisson's ratio $\nu = 0.25$. The panel is simulated by a linear elastic model, which is C30 concrete with a density of $\rho = 2.40$ g/cm^3, an elastic modulus of $E = 3.1 \times 10^4$ MPa, a Poisson's ratio of $\nu = 0.167$, and a concrete compressive strength $f_c = 27.6$ MPa. The tensile strength under static and seismic dynamic loads is calculated using the formula suggested by Raphael (1984).

Based on the above non-stationary stochastic ground motion generation method, 144 ground motions were generated. By comparing with the target value, the error of several control values is still within 10%, which meets the needs of stochastic

Table 3.1 Parameters of the generalized plasticity model of rockfill material

G_0	K_0	M_g	M_f	α_f	α_g	H_0	H_{U0}	m_s
1000	1400	1.8	1.38	0.45	0.4	1800	3000	0.5
m_s	m_v	m_l	m_u	r_d	γ_{DM}	γ_u	β_0	β_1
0.5	0.5	0.2	0.2	180	50	4	35	0.022

Table 3.2 Parameters of the generalized plasticity model of contact surface model

D_{s0}/kPa	D_{n0}/kPa	M_c	e_r	λ	a/kPa$^{0.5}$	b	c
1000	1500	0.88	0.4	0.091	224	0.06	3
a	r_d	k_m	M_f	k	H_0/kPa	f_h	t/m
0.65	0.2	0.6	0.65	0.5	8500	2	0.1

dynamics and probability analysis of CFRD. The vertical seismic acceleration is 2/3 of the horizontal seismic acceleration, and the horizontal bedrock peak acceleration is adjusted to 0.1–1.0 g, with an interval of 0.1 g. Input the ground motion and perform a series of finite element calculations. A total of 1440 working conditions need to be calculated for 10 seismic peak accelerations to obtain random dynamic information under the action of different intensities of ground motions. Then, based on the above numerical method, the probability density evolution equation is solved to obtain the probability information of seismic response of CFRD at each time. In the following, the influence of stochastic ground motion will be analyzed from the random dynamic response of dam body acceleration, dam body deformation, and face plate downslope stress, as well as the stochastic dynamic probability of these physical quantities.

3.4.2 Dam Acceleration

Figure 3.4 shows a cloud of the maximum horizontally oriented acceleration response based on a single sample and a cloud of the mean value of the response of 144 samples at 0.5 g seismic intensity. It can be seen that no matter based on the response obtained from a single sample or the average value of 144 sample responses, the acceleration response amplification in the crest area is the most obvious. In addition, the downstream dam slope is also a concentration of larger acceleration responses. However, the average distribution of the acceleration response is relatively regular and gentle, indicating that the stochastic seismic excitation has a huge impact on the maximum acceleration response of the dam body, and the stochastic seismic excitation makes the maximum acceleration response of each region tend to a constant value.

In order to further study the distribution law and influencing factors of the acceleration response of the dam body, Fig. 3.5 lists the distribution law of the maximum horizontal acceleration response with the dam height under the action of 0.2, 0.6 and 1.0 g ground motion intensity. We can see that the maximum horizontal acceleration responses caused by different ground motions are distributed differently along the dam height, but the overall trend is the same. A large turning point begins to appear at the dam height of 200 m, which is 0.8 H, and the magnification effect of the dam crest is abnormally obvious. It shows a strong "whipping effect" effect, and these characteristics are uniformly reflected in the distribution of the mean value along the

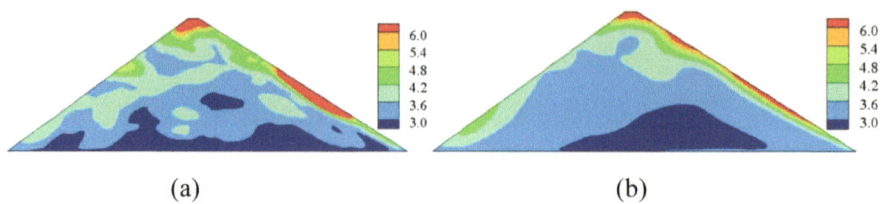

(a)   (b)

Fig. 3.4 Maximum horizontal acceleration response, **a** Single sample acceleration response, **b** 144 sample acceleration response mean

dam height. From the diagram of acceleration magnification (Fig. 3.5d), we can see that as the intensity of ground motion increases, the magnification decreases, and the acceleration response changes more dramatically with the increase of seismic intensity. Under the seismic intensity of 0.1 g, the average acceleration at the dam crest is approximately three times that at the dam base. At PGA (Peak Ground Acceleration) of 0.2, 0.6, and 1.0 g, the average acceleration at the dam crest is about 2.9, 2.5, and 2.4 times that at the dam base, respectively. The maximum horizontal acceleration along the dam height and amplification factors under other seismic intensities are listed in Table 3.3 (values in parentheses indicate amplification relative to the dam base). Interestingly, as the seismic intensity increases, the amplification effect decreases.

Figure 3.6 shows the time history of the horizontal acceleration response of the dam crest based on a single sample and the mean and standard deviation time history of 144 acceleration responses under the action of PGA = 0.5 g seismic intensity. There is a certain correlation between the time course of acceleration response and the time course of ground motion, and the mean value of acceleration of multiple ground motion tends to 0, indicating that the acceleration response has a strong variability, and the time course of the acceleration response varies a lot with different ground motion, so we should analyze the acceleration response of the CFRD based on the stochasticity of the ground motion. The time history of the standard deviation curve rises first, indicating that with the development of the nonlinear characteristics of the rockfill material, the variability of the acceleration response increases, and then decreases, partly because the variability of the acceleration response decreases with the decrease in the fluctuation amplitude of the time history of the ground motion. These changes prove that the acceleration response of CFRD is very sensitive to different ground motions, and shows that ground motion is a random process, so it is necessary to analyze the acceleration response from the perspective of random dynamics. Figure 3.7 shows the probabilistic evolution information of the dam top acceleration with time evolution under the action of PGA = 0.5 g seismic intensity, which can be obtained by the finite difference method in TVD format, including the probability density function at typical moments, the probability density evolution surface, and the probability density contour, which shows that the probability distribution of the acceleration response is irregular, and the probability density function surface is like the rolling peaks, which indicates that the acceleration response Rise

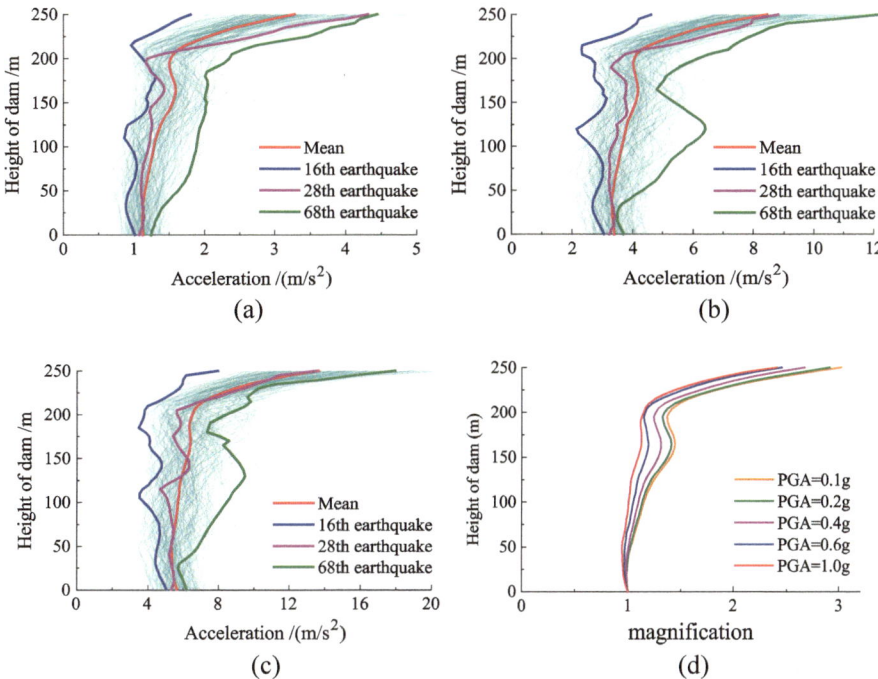

Fig. 3.5 Distribution of maximum horizontal acceleration along dam height, **a** PGA = 0.2 g, **b** PGA = 0.6 g, c PGA = 1.0 g, **d** different seismic intensities

Table 3.3 The mean of the maximum horizontal acceleration along the dam height and the magnification

Mean		PGA									
		0.1 g	0.2 g	0.3 g	0.4 g	0.5 g	0.6 g	0.7 g	0.8 g	0.9 g	1.0 g
Height of dam	0 m	0.56	1.13	1.69	2.26	2.81	3.38	3.98	4.54	5.13	5.69
	50 m	0.58	1.15	1.72	2.27	2.80	3.33	3.86	4.37	4.88	5.37
	100 m	0.65	1.28	1.89	2.48	3.06	3.63	4.18	4.70	5.23	5.72
	150 m	0.79	1.55	2.26	2.91	3.51	4.08	4.65	5.17	5.71	6.24
	200 m	0.78	1.52	2.20	2.85	3.44	4.03	4.63	5.19	5.87	6.54
	250 m	1.70 (3.04)	3.28 (2.90)	4.72 (2.79)	6.03 (2.67)	7.23 (2.57)	8.44 (2.50)	9.75 (2.45)	11.04 (2.43)	12.32 (2.40)	13.68 (2.40)

Note 0 m dam height indicates the bottom of the dam, and magnification is indicated in parentheses

and fall and evolve with time, also shows that the probability flows in the space state, which is better reflected in the probability density contour plots, and also shows that the acceleration response evolves with time with great variability.

Figure 3.8 shows the maximum horizontal acceleration exceeding probability at different dam heights under the 0.6 g seismic intensity, and the distribution of

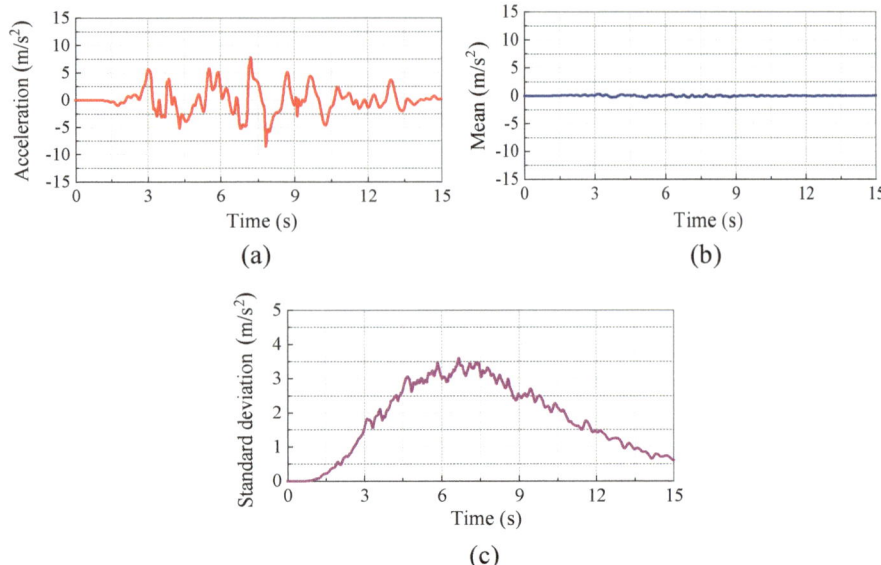

Fig. 3.6 Response, mean and standard deviation time-history of horizontal acceleration, **a** typical sample, **b** mean, **c** standard deviation

acceleration along the dam height under several typical exceeding probabilities, which can be obtained by constructing a virtual probability density evolution process combined with the SUPG format finite element method. It can be clearly seen that under different exceeding probabilities, the "whipping effect" effect is present above the dam height of 0.8 H, and the exceeding probability of the acceleration at the dam crest is much higher than that at other elevations, and the amplification effect is obvious. The 50% exceedance probability acceleration distribution is basically the same as the mean value, indicating that the generated stochastic ground motion has strong statistical laws and probabilistic significance.

Figure 3.9a shows the discrete point plots of the maximum horizontal acceleration at the top of the dam under the seismic intensity of 0.6 g. The point distributions have a certain degree of discretization, and the maximum value reaches about three times the minimum value, which indicates that the generated acceleration response under ground motion is statistically more significant. From the distributions of the mean maximum acceleration, 95% exceeding probability and 5% exceeding probability under different seismic intensities (Fig. 3.9b), it can be seen that the maximum horizontal acceleration basically has a linear distribution, and the trend of acceleration can be predicted more accurately by linear fitting; on the other hand, the maximum acceleration of the dam roof should be between 95 and 5% exceeding probability, when the PGA is 0.1, 0.2, 0.3, 0.4, 0.5, 0.6, 0.7, 0.8, 0.9 and 1.0 g, respectively. 0.3, 0.4, 0.5, 0.6, 0.7, 0.8, 0.9, and 1.0 g, the maximum horizontal acceleration response ranges from 0.9–2.61 m/s², 1.88–4.83 m/s², 2.99–6.70 m/s², 3.63–8.62 m/s², and

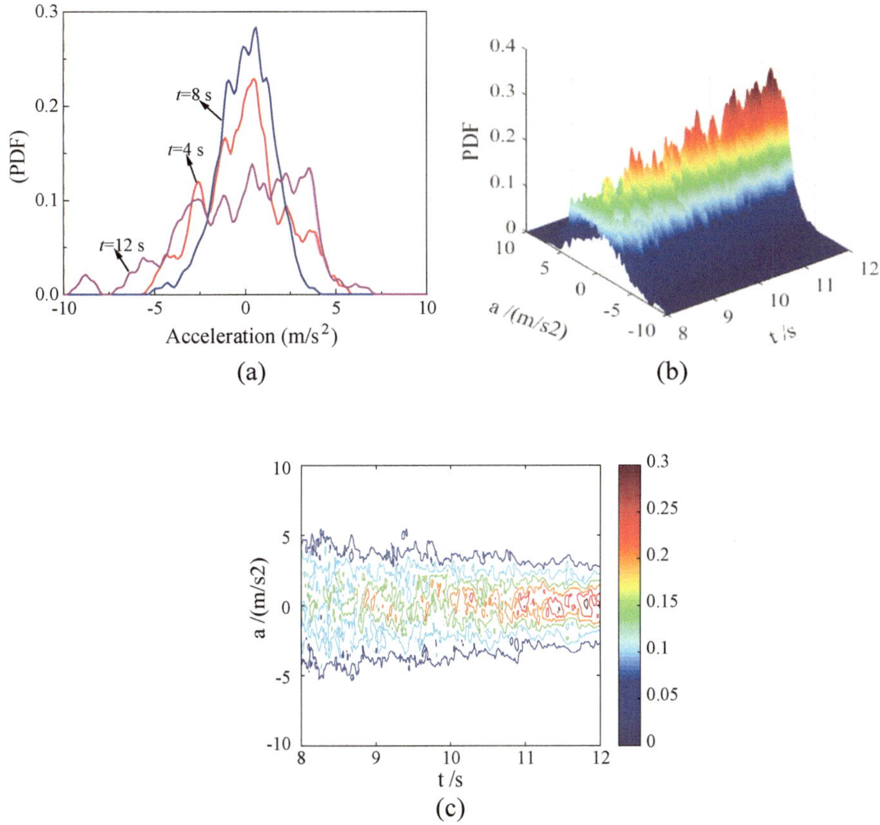

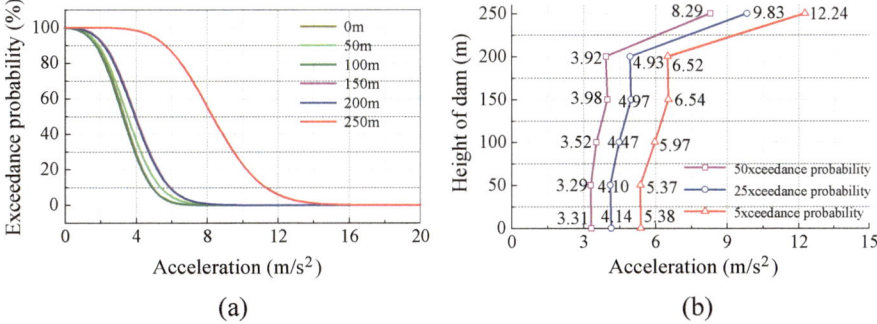

Fig. 3.7 Probability evolution information of horizontal acceleration in the dam crest, **a** PDF, **b** PDF evolution surfaces, **c** PDF contour

Fig. 3.8 The exceedance probability of the horizontal maximum acceleration and the distribution of acceleration along the dam height, **a** acceleration exceedance probability at different dam heights, **b** distribution of acceleration along the dam height for typical overtopping probability

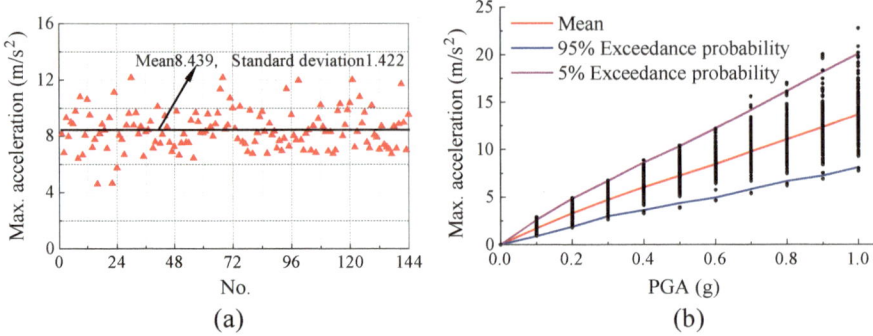

Fig. 3.9 Discrete point distribution of maximum acceleration response under different PGA. **a** Discrete point distribution of maximum acceleration (PGA = 0.6 g), **b** maximum acceleration at different seismic intensities

4.36–10.32 m/s² respectively, 4.93–12.24 m/s², 5.82–14.16 m/s², 6.67–16.15 m/s², 7.25–18.18 m/s², and 8.12–20.09 m/s², which can provide a reference for the seismic safety design of CFRD.

3.4.3 Dam Deformation

Figure 3.10 shows the horizontal residual deformation obtained based on a single sample at 0.5 g seismic intensity and the mean value of the response of 144 samples, while Fig. 3.11 shows the vertical residual deformation. It can be seen that both horizontal and vertical residual deformations occur at the top of the dam after the earthquake, whether based on the response of a single sample or the mean value of the response of 144 samples, and the distribution patterns of the response and the mean value of the response of a single sample are almost completely similar, which indicates that the effects of the ground motion on the deformation of the dam body of the various stripes of the law is basically similar.

Figure 3.12 shows the distribution pattern of horizontal residual deformation along the dam height for three seismic intensities of 0.1, 0.5 and 1.0 g. It can be seen that

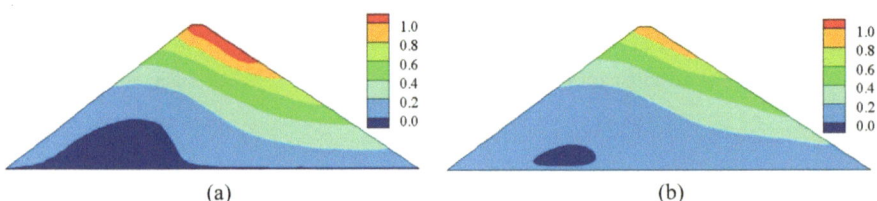

Fig. 3.10 Horizontal residual deformation, **a** single sample, **b** 144 samples

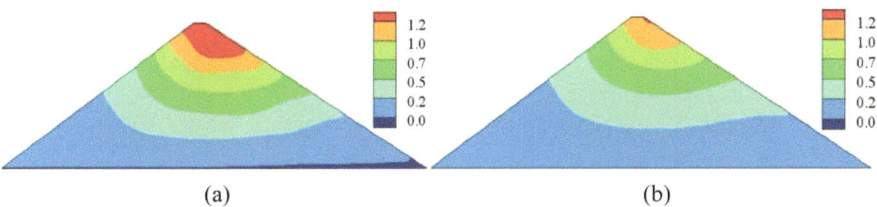

Fig. 3.11 Vertical residual deformation, **a** single sample, **b** 144 samples

the horizontal residual deformation caused by different ground motions has different distribution patterns along the dam height, but the trend is basically the same, it gradually increases along the dam height and reaches the maximum value at the top of the dam, which is consistent with the conclusion of the above mentioned distribution pattern of cloud diagrams, and the maximum horizontal residual deformation occurs at the top of the dam, but under the effect of strong earthquakes such as the 1.0 g earthquake the horizontal residual deformation does not appear completely at the top of the dam. Residual deformations caused by different ground motion are more discrete, and different seismic intensities also affect their distribution patterns, indicating that horizontal displacements are more sensitive to the effects of ground motion. Figure 3.13 shows the distribution of vertical residual deformation along the dam height under three seismic intensities of 0.1, 0.5 and 1.0 g, which is basically similar to that of horizontal residual deformation. The above analysis shows that the deformation of the dam body is more sensitive to the effect of ground motion, and the deformation dynamic response of CFRD should be analyzed from a stochastic point of view.

Figure 3.14 shows the horizontal displacement of the dam roof based on a single sample and the mean and standard deviation of 144 samples under PGA = 0.5 g seismic intensity, and Fig. 3.15 shows the vertical displacement. As seismic ground motion changes, dam crest deformation continuously increases and eventually stabilizes, exhibiting cumulative effects due to the plastic properties of the dam materials. However, it also shows slight fluctuations, indicating that the dam materials possess certain elastic properties and reloading process. The final stability of the mean value of the deformation is better, indicating that the generated stochastic ground motion is statistically better; the standard deviation generally increases, indicating that the variability of the deformation of the dam crest gradually increases with the development of the nonlinear behavior of the rockfill material; from the mean and standard deviation, the deformation is more sensitive to different ground motions. Figure 3.16 shows typical momentary probability density curves, probability density evolution surfaces, and probability density evolution contours for the horizontal displacement of the dam roof at 0.5 g seismic intensity, and Fig. 3.17 shows the probability information of vertical displacement. The probability distribution of deformation is not normal or lognormal as often assumed, but an irregular probability curve, the probability density surface shows the characteristics of peaks "high and low", and the

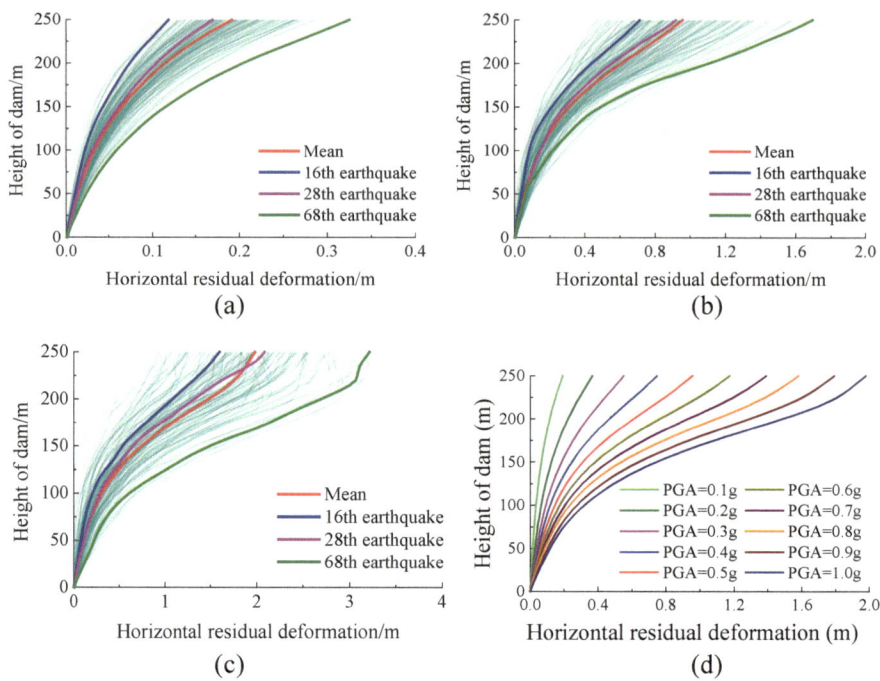

Fig. 3.12 Distribution of horizontal residual deformation along dam height, **a** PGA = 0.1 g, **b** PGA = 0.5 g, **c** PGA = 1.0 g, **d** different seismic intensities

contour lines flow like "water", which are the results of the irregular flow of probability in space. This is the result of the irregular flow in space, and the deformation evolves with time, which also reveals the transmission process of probability statistical information, and shows the sensitivity of deformation to different ground motion, and it is necessary to analyze the seismic deformation response of CFRD from the viewpoint of stochastic dynamics.

The exceeding probability of horizontal and vertical residual deformation of the dam roof is obtained by solving the constructed virtual stochastic process and generalized probability density evolution equations by the finite element method in SUPG format, as shown in Fig. 3.18 From the exceedance probability curves of horizontal and vertical residual deformations under different seismic intensities, the exceedance probability of each deformation value can be obtained to provide a basis for performance-based seismic design and safety evaluation of CFRD.

Figure 3.19a shows the discrete point distribution of the horizontal residual deformation of the dam roof under 0.5 g seismic intensity, the difference between the maximum and minimum values is large, the maximum value reaches about 5 times of the minimum value, but it is more concentrated near the mean value, which indicates that the generated horizontal displacement of the dam body under the action of

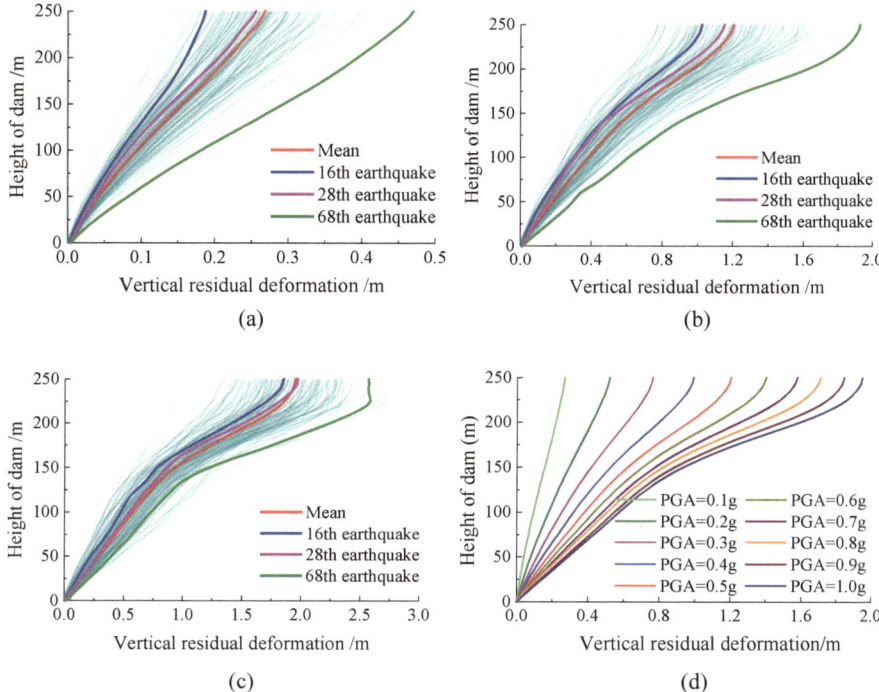

Fig. 3.13 Distribution of vertical residual deformation along dam height, **a** PGA = 0.1 g, **b** PGA = 0.5 g, **c** PGA = 1.0 g, **d** different seismic intensities

the ground motion is more statistically significant. From the discrete point distribution of the horizontal residual deformation of the dam roof under different seismic intensities and the distribution curves of the mean value, 50% exceeding probability, 95% exceeding probability, and 5% exceeding probability (Fig. 3.19b), the horizontal residual deformation basically shows a linear distribution, and the trend of the horizontal residual deformation can be predicted more accurately by the fitting of the formula; and it is obvious that the curves of the mean value and 50% exceeding probability almost coincide, indicating that the generated random ground motion has a stronger statistical regularity. almost coincide, indicating that the generated random ground motion has a strong statistical regularity. On the other hand, the post-earthquake horizontal deformation of the dam roof should be distributed between 95 and 5% beyond probability, with the variation ranging from 0.09–0.29 m, 0.18–0.58 m, 0.27–0.89 m when the PGA is 0.1, 0.2, 0.3, 0.4, 0.5, 0.6, 0.7, 0.8, 0.9 and 1.0 g, respectively, 0.37–1.21 m, 0.44–1.58 m, 0.56–1.92 m, 0.65–2.25 m, 0.68–2.58 m, 0.84–2.87 m, and 0.91–3.20 m, respectively, which can provide a reference for the seismic safety design of CFRD. Figure 3.20a shows the distribution point plot of the vertical residual deformation of the dam roof under 0.5 g seismic intensity, and the maximum value is about four times of the minimum value; from the discrete point distribution of the vertical residual deformation under different seismic

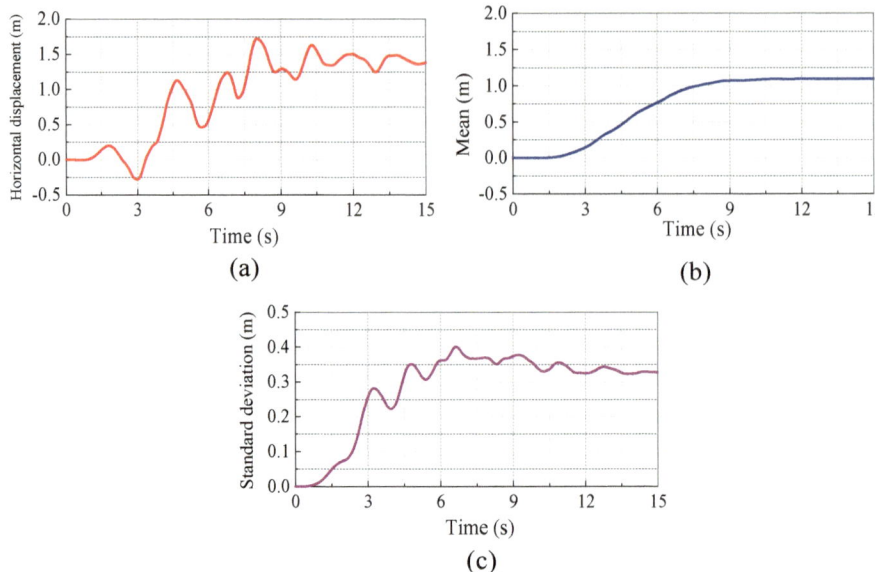

Fig. 3.14 Mean and standard deviation time history of horizontal displacement, **a** typical sample, **b** mean, **c** standard deviation

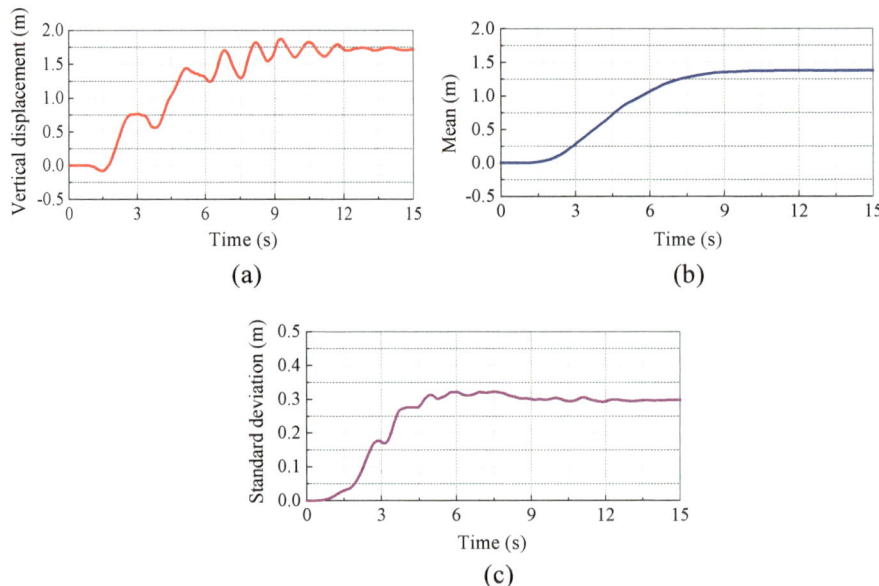

Fig. 3.15 Mean and standard deviation time history of vertical displacement, **a** typical sample, **b** mean, **c** standard deviation

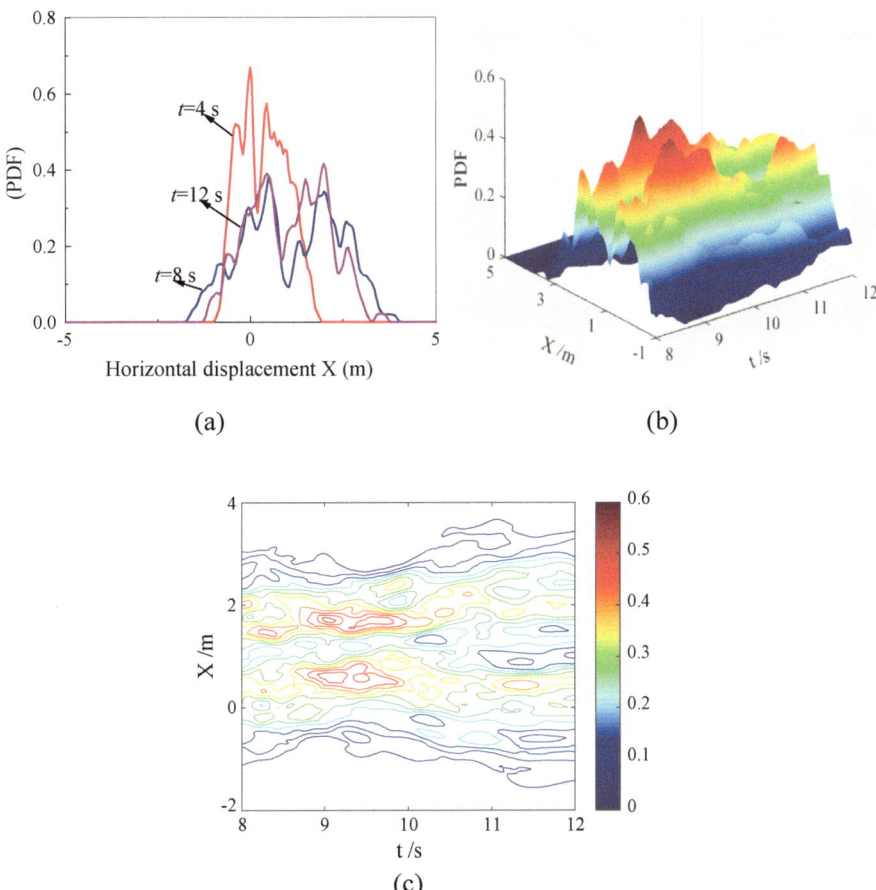

Fig. 3.16 Probability evolution information of horizontal displacement in the dam crest **a**, PDF, **b** PDF evolution surface, **c** PDF contours

intensities and the distribution curves of the mean, 50% exceeding probability, 95% exceeding probability, and 5% exceeding probability (Fig. 3.20b), with the increase of the seismic intensity, as the stacked rock material becomes more and more dense resulting in the change of distribution curve gradually becomes slower; when the PGA varies from 0.1 to 1.0 g, the range of post seismic vertical deformation of the dam roof should be 0.14–0.41 m, 0.28–0.79 m, 0.40–1.17 m, 0.53–1.51 m, 0.67–1.79 m, 0.79–2.07 m, 0.92–2.28 m, 0.94–2.52 m, 1.07–2.66 m, and 1.16–2.77 m, which provide references for the dam seismic design and ultimate seismic capacity analysis. Table 3.4 lists the horizontal and vertical residual deformations of the dam roof at 50, 95 and 5% exceeding probability.

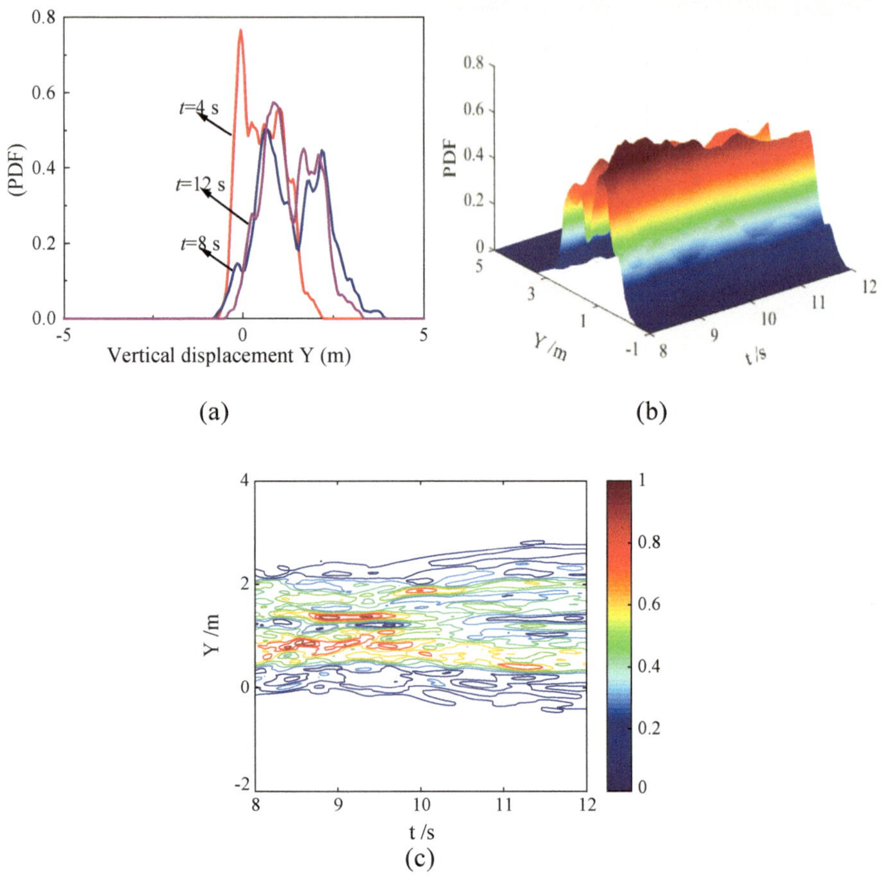

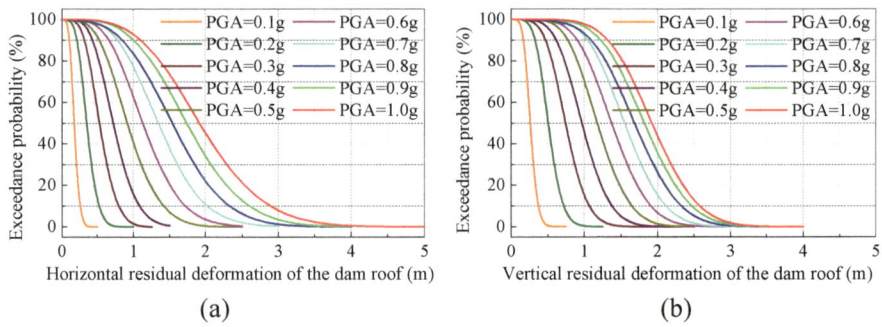

Fig. 3.17 Probability evolution information of vertical displacement in the dam crest, **a** PDF, **b** PDF evolution surface, **c** PDF contours

Fig. 3.18 Exceedance probability of horizontal and vertical residual deformation under different PGA. **a** horizontal residual deformation exceeding probability, **b** vertical residual deformation beyond probability

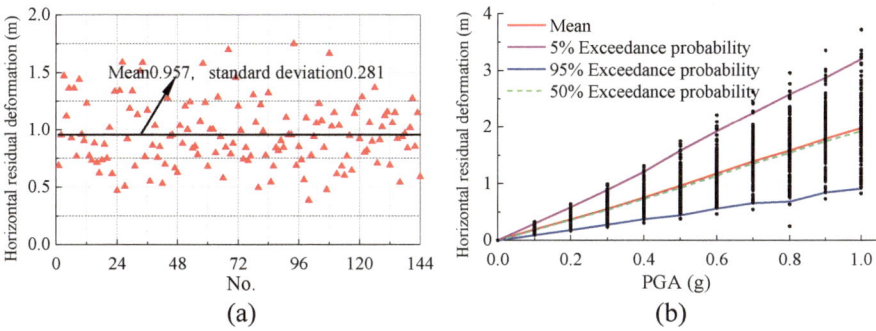

Fig. 3.19 Horizontal residual deformation of dam crest under different PGA, **a** discrete point distribution of horizontal residual deformation (pga = 0.5 g), **b** horizontal residual deformation at different seismic intensities

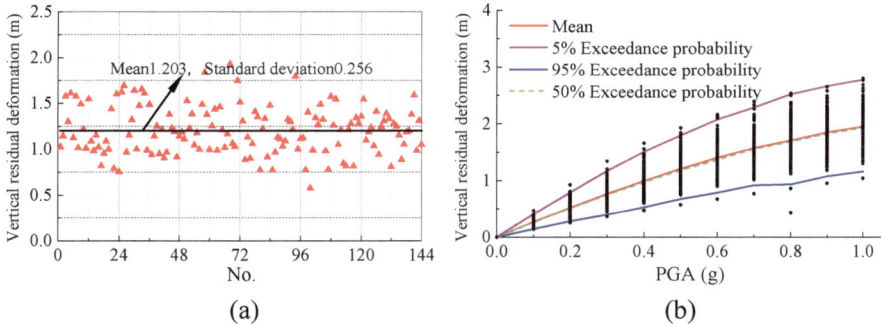

Fig. 3.20 Vertical residual deformation of dam crest under different PGA, **a** discrete point distribution of vertical residual deformation (PGA = 0.5 g), **b** vertical residual deformation under different seismic intensities

Table 3.4 The horizontal and vertical residual deformation of dam crest based on different exceedance probability under different PGA

Exceedance probability situation		PGA									
		0.1 g	0.2 g	0.3 g	0.4 g	0.5 g	0.6 g	0.7 g	0.8 g	0.9 g	1.0 g
Horizontal displacement (m)	Mean	0.192	0.367	0.553	0.749	0.957	1.174	1.388	1.579	1.790	1.979
	5%	0.292	0.584	0.886	1.205	1.580	1.920	2.246	2.577	2.872	3.20
	50%	0.178	0.355	0.534	0.725	0.920	1.131	1.346	1.539	1.741	1.921
	95%	0.089	0.179	0.273	0.370	0.440	0.560	0.653	0.683	0.838	0.910
Vertical displacement (m)	Mean	0.268	0.518	0.762	0.990	1.203	1.401	1.574	1.707	1.845	1.948
	5%	0.408	0.788	1.171	1.505	1.791	2.071	2.284	2.518	2.662	2.774
	50%	0.262	0.506	0.744	0.965	1.178	1.362	1.551	1.686	1.823	1.927
	95%	0.142	0.278	0.396	0.525	0.674	0.786	0.917	0.935	1.073	1.160

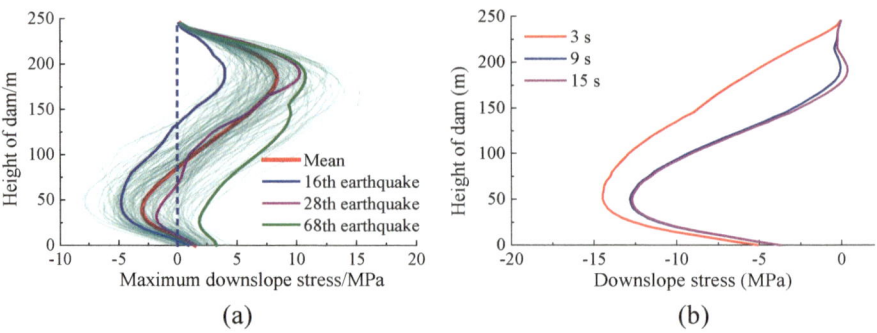

Fig. 3.21 The maximum stress distribution along the dam height, **a** maximum downhill stress, **b** mean value of downslope stress at different moments

3.4.4 Panel Stress

Figure 3.21 shows the distribution pattern of panel downslope stresses along the dam height of 0.4 g seismic intensity (where tensile stresses are positive and compressive stresses are negative). The distribution pattern of stress response along the dam height caused by different ground motions is different, but the trend is basically the same; the average value of the downslope tensile stress is mainly concentrated in the range of 100–250 m of the dam height, and reaches the maximum value around 0.75 H of the dam height. From the distribution of stress averages along the dam height at 3, 9 and 15 s moments, the panels are mainly subjected to compressive stresses and are more compressive at the bottom of the panels, which are generally in an extruded state. The above analysis shows that, from a random perspective, the values and distribution patterns of panel stress response caused by different ground motions are quite different, and the stress response analysis of individual ground motions cannot effectively evaluate the change rule, reflecting the necessity of analysis based on the randomness of ground motions.

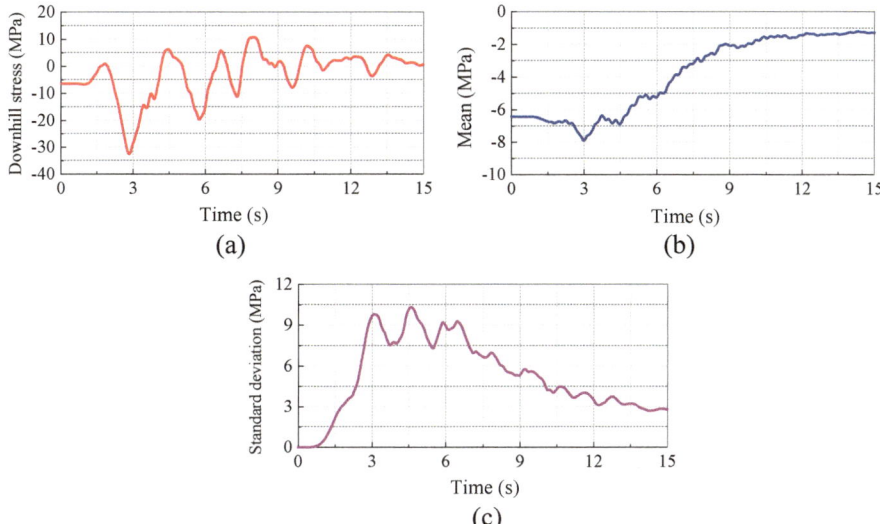

Fig. 3.22 Stress response, mean and standard deviation time history, **a** typical sample, **b** mean, **c** standard deviation

As a representative, Fig. 3.22 gives the stress time course based on a single sample and the mean and standard deviation time course of the stress of 144 samples under 0.4 g seismic intensity. The ups and downs of the stresses show the open and closed state of the concrete; the characteristics of the mean stresses with time show that the panels are mainly in the compression state from the stochastic dynamic analysis considerations; the different ground motions have a great influence on the stress changes of the panels; the standard deviation of the stresses increases firstly and then decreases with time, which demonstrates the opening and closing of the concrete's linear properties, which is caused by the variation of the strength of the ground motions. Figure 3.23 illustrates the state of evolution of the stress probability density with time, which shows the sensitivity of the panel stresses to different ground motions.

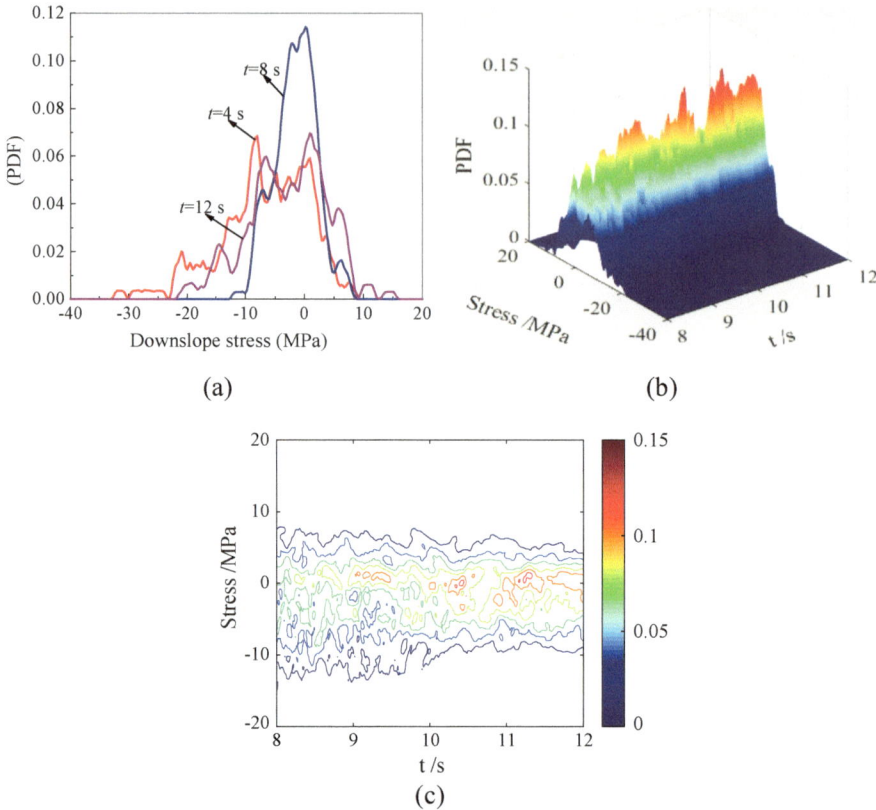

Fig. 3.23 Probability density evolution information of stress, **a** typical PDF curves, **b** PDF evolution surface, **c** PDF contours

3.5 Conclusion

In this chapter, the uncertainty of ground motion input is fully considered, and the random ground motion generation method, generalized probability density evolution method, probabilistic analysis of susceptibility, and elastic–plastic analysis of panel rockfill dams are jointly applied to reveal the stochastic dynamic response and probabilistic rule of change of the seismic process and after earthquake of CFRD based on the physical quantities of acceleration, deformation, and panel stresses of the dam, and the distribution change ranges of the physical quantities are proposed from the stochastic dynamics and probability points of view. Finally, considering the two evaluation aspects of dam body deformation and panel impermeable body safety, appropriate performance indexes are selected, different performance level classification standards are initially proposed, and the performance safety evaluation framework of multi-seismic intensity-multi-performance target-exceeding probability is established. The main work and conclusions are as follows:

(1) From the perspective of stochastic dynamics, it is revealed that the maximum horizontal acceleration distribution of the dam body is mainly concentrated in the top of the dam and the downstream slope area, and above 0.8 H (H is the height of the dam) shows a strong "whip-sheath" effect, and the amplification multiples of the top of the dam are different under the action of different ground motion intensities; through the stochastic dynamics analysis and probabilistic analysis of strong statistical significance, it is shown that the maximum value is three times or even five times of the minimum value. Probability analysis shows that the randomness of ground motion has a large impact on the acceleration, deformation and panel stress response of the dam body, and the maximum value is three times or even five times of the minimum value, so it is necessary to comprehensively evaluate the seismic capacity of CFRD from the random point of view of ground motion; reveals the characteristics of the probability of the response of various physical quantities of the spatial irregularities in the flow of the characteristics of the probability of the distribution of the response of the response of the response of the response of the response of each physical quantity in the space is not the regular probability of distribution such as the characteristics of the normal distribution; based on the 5, 50% and Based on 5, 50 and 95% exceeding probability, the range of calculated values of different physical quantities is suggested, which provides a reference for the numerical calculation results and the analysis of ultimate seismic capacity of CFRD.

Uncertainty in rockfill material parameters also has an impact on the establishment of a unified performance safety evaluation standard for CFRD, so the stochastic dynamic response and probabilistic analysis results of CFRD should be further investigated on the basis of full consideration of ground motion stochasticity, material parameter uncertainty including ground motion-material parameter coupled stochasticity, and then the performance-based seismic safety evaluation framework should be improved.

References

Deeks AJ, Randolph MF (1994) Axisymmetric time-domain transmitting boundaries. J Eng Mech 120(1):25–42

Design code for concrete face rockfill Dams SL228-2013. China Water & Power Press, Beijing

Feng TG, Yang G (2010) Analysis on collapse-proof effectiveness of unseating-prevention device under earthquake sequences. J Catastrophol 25(3):42–48

Kong XJ, Liu JM, Zou DG et al (2014) Experimental study of particle breakage of Zipingpu rockfill material. Rock Soil Mechan 35(1):35–40

Liu HB, Ling HI (2008) Constitutive description of interface behavior including cyclic loading and particle breakage within the framework of critical state soil mechanics. Int J Numer Anal Meth Geomech 32(12):1495–1514

Liu JB, Lv YD (1998) A direct method for analysis of dynamic soil-structure interaction. Chin Civil Eng J 31(3):55–64

Liu JB, Wang ZY, Du XL et al (2005) Three-dimensional visco-elastic artificial boundaries in time domain for wave motion problems. Eng Mechan 22(6):46–51

Liu JB, Gu Y, Du YX (2006) Consistent viscous-spring artificial boundaries and viscous-spring boundary elements. Chin J Geotech Eng 28(9):1070–1075

Liu J, Liu B, Kong XJ (2012) Estimation of earthquake-induced crest settlements of earth and rock-fill dams. J Hydroelectr Eng 31(2):183–191

Liu JM, Zou DG, Kong XJ (2014) A three-dimensional state-dependent model of soil–structure interface for monotonic and cyclic loadings. Comput Geotech 61:166–177

Liu JM, Kong XJ, Zou DG (2015) Effects of particle breakage due to vibration compaction on the deformation behavior of rockfill dam. J Hydraul Eng 46(8):934–942

Lysmer J, Kuhlemeyer RL (1969) Finite dynamic model for infinite media. J Eng Mech Div 95(4):859–877

Raphael JM (1984) Tensile strength of concrete. J Proc 81(2):158–165

Xu B, Zou DG, Liu HB (2012) Three-dimensional simulation of the construction process of the Zipingpu concrete face rockfill dam based on a generalized plasticity model. Comput Geotech 43:143–154

Zhou Y (2012) Seismic damage analysis of Zipingpu panel rockfill dam and panel seismic countermeasures for the Wenchuan earthquake. Dalian University of Technology, Dalian

Zou DG, Xu B, Kong XJ et al (2013) Numerical simulation of the seismic response of the Zipingpu concrete face rockfill dam during the Wenchuan earthquake based on a generalized plasticity model. Comput Geotech 49:111–122

Chapter 4
Stochastic Dynamic Analysis for High CFRD Considering Uncertainties of Material Parameters

4.1 Introduction

Due to the uncertainty of the source of dam construction materials for CFRD, the material model parameters should be uncertain values, so the traditional deterministic seismic safety evaluation method is difficult to objectively evaluate the CFRD, especially difficult to analyze from the perspective of the probability of performance, and there are only a small number of studies that consider the effect of uncertainty in the dam construction materials, and there is virtually no literature to consider the effect of the randomness of elasticity-plasticity parameters of the dam construction materials on the seismic There are only a few studies considering the effect of uncertainty of dam construction materials, and almost no literature considering the effect of randomness of elastic–plastic parameters of dam construction materials on the seismic response, while the effect of parameter uncertainty on the seismic response of CFRD cannot be ignored. Therefore, the stochastic dynamic and probabilistic analysis of CFRD based on the uncertainty of material parameters can be an effective supplement to the deterministic analysis method, which is of great significance, and will become an important means to scientifically and quantitatively study and ensure the seismic safety of CFRD, and to provide a basis for the performance-based seismic safety evaluation of CFRD.

4.2 Determination of Random Variables for Elastic–Plastic Material Parameters of CFRD

From Sect. 3.2.1, it can be seen that there are 17 parameters in the generalized plasticity model of rockpile material, theoretically, they all have some influence on the deformation, and all the parameters should be considered as random variables, but there is a certain degree of correlation between them, and there are too many

B. Xu and R. Pang, *Stochastic Dynamic Response Analysis and Performance-Based Seismic Safety Evaluation for High Concrete Faced Rockfill Dams*, Hydroscience and Engineering, https://doi.org/10.1007/978-981-97-7198-1_4

random variables which often make the result of the analysis is too complex and inaccurate, and many of the parameters have little influence on the deformation of the dam and can be ignored in the stochastic dynamic analysis and reliability analysis. Many parameters have very little effect on dam deformation, which can be ignored in stochastic dynamic analysis and reliability analysis. Therefore, in this section, we find the key factors and determine the random variables by analyzing the sensitivity of material parameters to dam deformation.

The finite element model and material parameter information in this section is essentially the same as in Sect. 3.4.1 and is adjusted based on the elastic–plastic parameters in the table. The following sensitivity analyses will be carried out in terms of the three aspects of the generalized plasticity parameter determination: elasticity-dependent modulus, loaded plasticity-dependent modulus, and unloaded plasticity-dependent modulus parameters.

(1) Sensitivity analysis of elastic correlation modulus parameters

There are four elastic modulus parameters in the generalized plasticity model, namely, G_0, K_0, m_s, m_v. The results of the dynamic response of the dam corresponding to the parameter changes are shown in Table 4.1, and the trend plot of the dynamic response of the dam with the parameter changes is shown in Fig. 4.1. From the figure and table, it can be seen that: the sensitivity of the vertical deformation of the dam roof to this group of parameters decreases in the order of m_s, m_v, G_0, K_0, in which the parameters m_s and m_v are positively correlated with the deformation, and the parameter G_0 is negatively correlated with the deformation. The slope of the parameter m_s curve on the trend graph is the largest, that is, it has the greatest influence, and its increase by 30%, the deformation of the dam body increases by 9.19%. A positive or negative rate of change of 5% or more is regarded as a significant change, so m_s in the elastic correlation modulus parameter is regarded as a random variable.

(2) Sensitivity analysis of loaded plasticity-related modulus parameters

There are 10 plastic modulus parameters loaded in the generalized plasticity model, namely, M_g, M_f, α_f, α_g, H_0, m_l, r_d, γ_{DM}, β_0, β_1. The results of the dynamic response of the dam corresponding to the parameter changes are shown in Table 4.2, and the trend graph of the dynamic response of the dam with the parameter changes is shown in Fig. 4.2. From the figure and table, it can be seen that: the parameters α_g, γ_{DM} have a small influence (within 1.78%) and can be considered as quantitative, and the influence of the rest of the parameters on the maximum vertical deformation of the dam body decreases according to the order of M_f, H_0, r_d, α_f, β_1, β_0, m_l, M_g. Among them, the effect of α_f is positively correlated, and the effects of M_f, H_0, r_d, β_1, β_0 are negatively correlated. When M_f increases by 30%, the change in vertical deformation induced is -33.39%, which is much larger than the change induced by the remaining random variables. The effects of the remaining parameters are relatively small overall, with H_0 increasing from 20 to 30% causing a change of 2.95%, which is only 1/3 of that of M_f in the same case. loaded plastic correlation modulus parameters M_g, M_f, α_f, H_0, r_d, β_0, β_1 are considered as random variables.

Table 4.1 The variation of elastic modulus and its corresponding deformation

Parameter	Change (%)	Vertical deformation		Parameter	Change (%)	Vertical deformation	
		Mix (m)	Rate (%)			Mix (m)	Rate (%)
G_0	− 30	1.470	3.16	K_0	− 30	1.385	− 2.81
	− 20	1.455	2.06		− 20	1.415	− 0.73
	− 10	1.434	0.62		− 10	1.426	0.04
	0	1.425	0		0	1.425	0
	10	1.420	− 0.37		10	1.408	− 1.21
	20	1.425	− 0.04		20	1.400	− 1.75
	30	1.433	0.58		30	1.399	− 1.85
m_s	− 30	1.367	− 4.09	m_v	− 30	1.355	− 4.95
	− 20	1.383	− 2.95		− 20	1.384	− 2.90
	− 10	1.394	− 2.20		− 10	1.411	− 0.96
	0	1.425	0		0	1.425	0
	10	1.463	2.64		10	1.431	0.41
	20	1.516	6.37		20	1.434	0.60
	30	1.556	9.19		30	1.443	1.24

Fig. 4.1 The relationship between variation of elastic modulus and vertical deformation

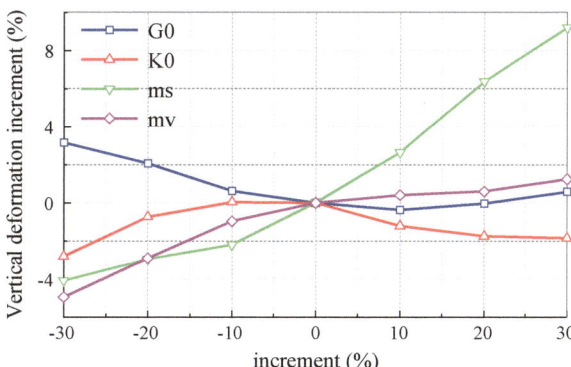

(3) Sensitivity analysis of unloaded plasticity-related modulus parameters

There are three unloaded plastic modulus parameters in the generalized plasticity model, namely H_{U0}, m_u, γ_u. According to Table 4.3, when the three parameters are varied within the range of ± 30%, the dynamic response of the dam does not change much, and the range of variations is basically within 1%, so the model can be disregarded regarding the parameter's sensitivity analysis.

In summary, m_s, M_g, M_f, α_f, H_0, γ_d, β_0, β_1 are regarded as random variables in the study of this chapter, a total of eight random parameters, and other variables are

Table 4.2 The variation of loading plasticity modulus and its corresponding deformation

Parameter	Change (%)	Vertical deformation		Parameter	Change (%)	Vertical deformation	
		Mix (m)	Rate (%)			Mix (m)	Rate (%)
M_g	− 30	1.548	8.62	M_f	− 30	1.656	16.22
	− 20	1.425	0.01		− 20	1.562	9.61
	− 10	1.409	− 1.12		− 10	1.494	4.86
	0	1.425	0		0	1.425	0
	10	1.449	1.71		10	1.232	− 13.56
	20	1.452	1.85		20	1.071	− 24.83
	30	1.471	3.23		30	0.949	− 33.39
α_f	− 30	1.216	− 14.70	α_g	− 30	1.400	− 1.78
	− 20	1.312	− 7.92		− 20	1.409	− 1.15
	− 10	1.352	− 5.15		− 10	1.417	− 0.57
	0	1.425	0		0	1.425	0
	10	1.451	1.84		10	1.434	0.59
	20	1.461	2.55		20	1.442	1.17
	30	1.481	3.90		30	1.448	1.61
H_0	− 30	1.595	11.94	m_l	− 30	1.479	3.80
	− 20	1.537	7.86		− 20	1.462	2.59
	− 10	1.479	3.79		− 10	1.443	1.27
	0	1.425	0		0	1.425	0
	10	1.376	− 3.47		10	1.408	− 1.18
	20	1.326	− 6.93		20	1.386	− 2.74
	30	1.284	− 9.88		30	1.367	− 4.09
r_d	− 30	1.577	10.62	γ_{DM}	− 30	1.450	1.77
	− 20	1.523	6.88		− 20	1.440	1.07
	− 10	1.473	3.39		− 10	1.432	0.49
	0	1.425	0		0	1.425	0
	10	1.382	− 3.00		10	1.419	− 0.41
	20	1.340	− 6.01		20	1.414	− 0.76
	30	1.302	− 8.65		30	1.410	− 1.06
β_0	− 30	1.503	5.47	β_1	− 30	1.546	8.50
	− 20	1.477	3.66		− 20	1.492	4.70
	− 10	1.447	1.54		− 10	1.459	2.39
	0	1.425	0		0	1.425	0
	10	1.399	− 1.83		10	1.394	− 2.20
	20	1.382	− 3.04		20	1.362	− 4.42
	30	1.362	− 4.41		30	1.317	− 7.58

Fig. 4.2 The relationship between variation of loading plasticity modulus and deformation

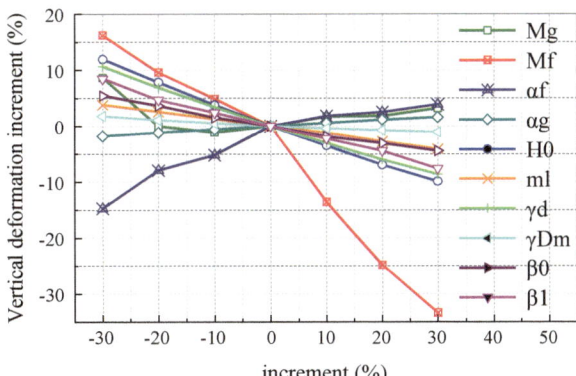

Table 4.3 The variation of unloading plasticity modulus and its corresponding deformation

Parameter	Change (%)	Vertical deformation		Parameter	Change (%)	Vertical deformation	
		Mix (m)	Rate (%)			Mix (m)	Rate (%)
H_{U0}	− 30	1.413	− 0.84	m_u	− 30	1.423	− 0.13
	− 20	1.419	− 0.42		− 20	1.424	− 0.06
	− 10	1.423	− 0.15		− 10	1.424	− 0.06
	0	1.425	0		0	1.425	0
	10	1.428	0.22		10	1.426	0.06
	20	1.430	0.34		20	1.426	0.06
	30	1.432	0.48		30	1.427	0.13
γ_u	− 30	1.407	− 1.26				
	− 20	1.415	− 0.74				
	− 10	1.421	− 0.28				
	0	1.425	0				
	10	1.429	0.29				
	20	1.432	0.48				
	30	1.434	0.62				

regarded as fixed values, and due to the current generalized plasticity model parameter information is not enough accumulated, the correlation between the parameters is not considered for the time being, and the following is only qualitatively and quantitatively a preliminary investigation of the effect of the material parameter stochasticity on the stochastic dynamic response of the high-panel rockfill dams and the probability results.

Table 4.4 Random parameters of the generalized plasticity model of rockfill material

Parameter	G_0	K_0	M_g	M_f	α_f	α_g	H_0	H_{U0}	m_s
Value	1000	1400	1.8	1.38	0.45	0.4	1800	3000	0.5
Coefficient of variation	–	–	0.1	0.1	0.1	–	0.1	–	0.1
Parameter	m_v	m_l	m_u	r_d	γ_{DM}	γ_u	β_0	β_1	
Value	0.5	0.2	0.2	180	50	4	35	0.022	
Coefficient of variation	–	–	–	0.1	–	–	0.1	0.1	

4.3 Stochastic Dynamic Response and Probabilistic Analysis of High CFRD

4.3.1 Basic Information

The finite element model and material parameter information and loading conditions used in this section are basically the same as those in Sect. 3.4.1, with peak ground motion acceleration PGA = 0.5 g. The ground motion time history of the intermediate value of the deformation response in Sect. 3.4 above is used. Random information on the material parameters is shown in Table 4.4, assuming normal or lognormal distributions, and comparing the effects of different distribution types. Considering the "3σ" criterion and the actual situation in numerical calculation, the coefficient of variation of each random material parameter is assumed to be 0.1. 144 sets of material parameter samples are generated based on the above GF-deviation optimization method, and then a series of finite element dynamics calculations are carried out, which are coupled with the generalized probability density evolution method mentioned above, to obtain the acceleration and probability information of the high-panel stacked rock dam and the deformation and panel stress response of the dam body. The stochastic dynamic and probabilistic information of the acceleration, deformation and panel stress response of the dam body also provides a reference for the performance safety evaluation considering the stochasticity of the material parameters.

4.3.2 Dam Acceleration

Figure 4.3 shows the maximum horizontal acceleration response cloud obtained based on the mean material parameters and the mean value cloud of all sample responses. It can be seen that the acceleration response amplification in the dam top region is the most obvious, and the downstream dam slope is also a concentrated area of larger acceleration response; the mean value of the response and the sample response of a single group of material parameters are basically the same

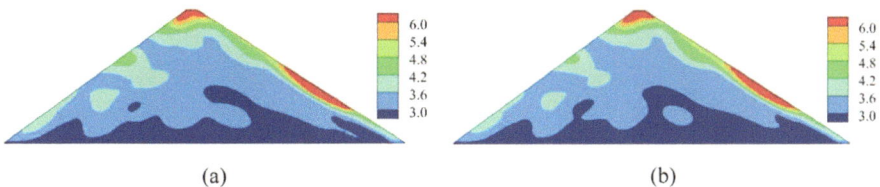

Fig. 4.3 Maximum horizontal acceleration response (**a**) single group sample (**b**) mean

distribution law except for the numerical difference, showing different characteristics from the random ground motion excitation, which indicates that the material parameter stochasticity is different from the seismic stochastic response law induced by the randomness of the ground motion, and therefore, the consideration of the material parameter Therefore, it is necessary to consider the randomness of material parameters.

From the distribution law of maximum horizontal acceleration along the dam height in Fig. 4.4, it can be seen that the distribution law of maximum horizontal acceleration response along the dam height caused by different material parameter samples varies, but the general trend is basically the same, all of them begin to show a large turnaround at 200 m, i.e., at the dam height of 0.8H; however, the distribution law shows different distribution characteristics from the ground shaking stochasticity, and does not have ground motion stochasticity However, the distribution pattern shows different distribution characteristics from the ground motion randomness, and there is no large dispersion of seismic response caused by ground motion randomness; when considering the normal distribution of material parameters, the maximum value of acceleration at the top of the dam is 10.6 m/s^2, and the minimum value is 5.6 m/s^2; comparing the acceleration response by considering the normal distribution of material parameters and the lognormal distribution, the difference between the two is not particularly obvious, and the mean response at the top of the dam is normally distributed as 8.2 m/s^2, and the lognormal distribution is 8.6 m/s^2.

Figure 4.5 shows the acceleration response of the dam top obtained based on the mean value of material parameters and the mean and standard deviation time history of all sample responses. It can be seen that the acceleration response caused by material parameter randomness and ground motion randomness show different patterns of change, indicating the necessity of considering material parameter randomness; the drastic change of the standard deviation indicates that the material parameter randomness has a more drastic effect on the acceleration response under the influence of the ground motion; and the consideration of the material parameter normal distribution and the lognormal distribution does not have a great influence on the time history of the acceleration response.

From the discrete point distribution plot of the maximum acceleration response at the top of the dam and the exceeding probability plot demonstrated in Fig. 4.6, the maximum response of acceleration is different with different material parameters, and there is a certain difference. For normal distribution, the maximum value is

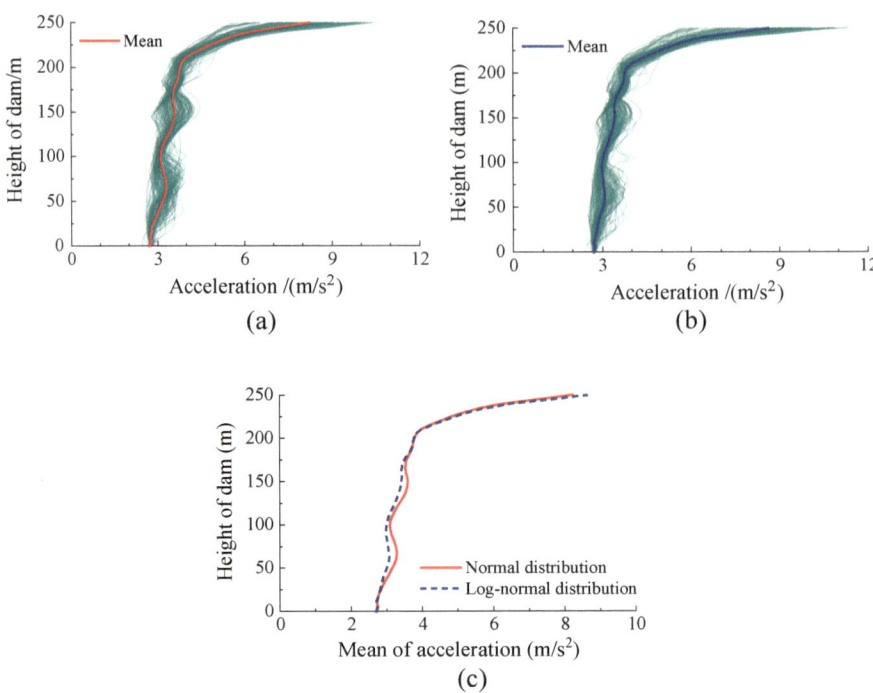

Fig. 4.4 Distribution of maximum horizontal acceleration along dam height, **a** normal distribution of material parameters, **b** log-normal distribution of material parameters, **c** comparison of mean values of acceleration response due to different distribution types

10.6 m/s² and the minimum value is 5.6 m/s², with a difference of about 1 times; the value of 95% exceeding probability is 5.9 m/s² and the value of 5% exceeding probability is 10.5 m/s², with a difference of about 0.8 times; for the lognormal distribution, the maximum value is 11.3 m/s² and the minimum value is 5.6 m/s², with the mean value and the value of the maximum acceleration response under each exceeding probability The difference is also small, within 5%. Therefore, the effect of the randomness of the material parameters on the acceleration response should be considered, but the effect of the type of distribution of the material parameters on the acceleration response can be disregarded.

4.3.3 Dam Deformation

Figure 4.7 shows the horizontal and vertical residual deformation responses obtained based on the mean material parameters and the mean distribution of the responses for all samples. It can be seen that the maximum horizontal and vertical residual deformations occur at the top of the dam considering both single and multiple sample

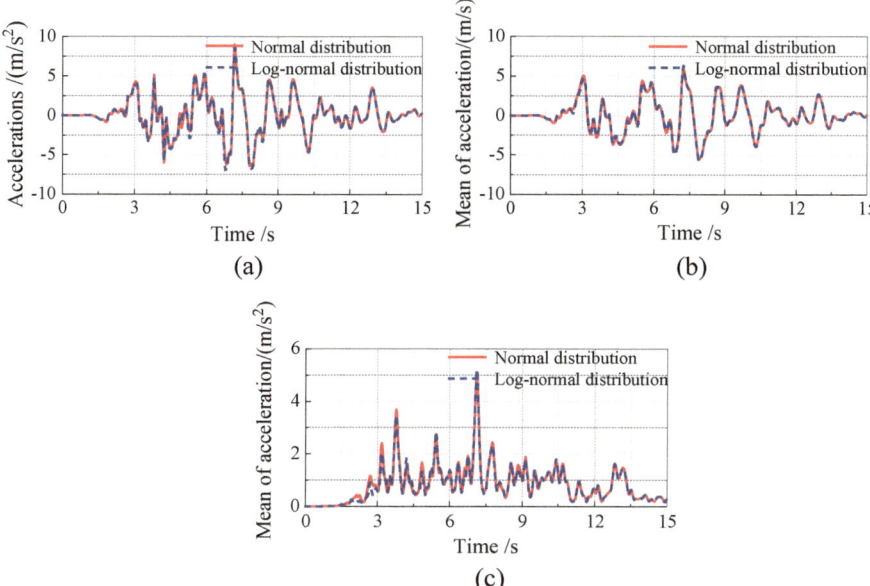

Fig. 4.5 Horizontal acceleration time history and its mean and standard deviation, **a** typical sample, **b** mean, **c** standard deviation

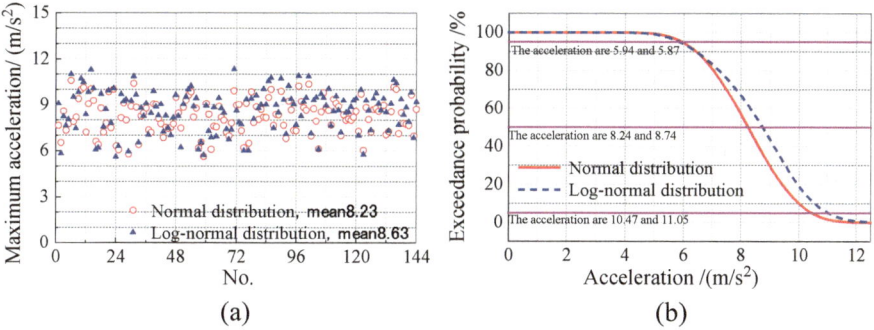

Fig. 4.6 Discrete point distribution of maximum acceleration and exceedance probability, **a** discrete point distribution of maximum acceleration, **b** seismic fragility curves of crest maximum acceleration

means, and that the responses based on a single set of samples are almost exactly similar to the law of distribution of the means, differing only numerically.

Figure 4.8 shows the distribution law of horizontal residual deformation along the dam height. It can be seen that the horizontal residual deformation along the dam height caused by different material parameters has different distribution laws, but the trend is basically the same, along the height of the dam gradually increases, the dam top reaches the maximum value, which is consistent with the above distribution law

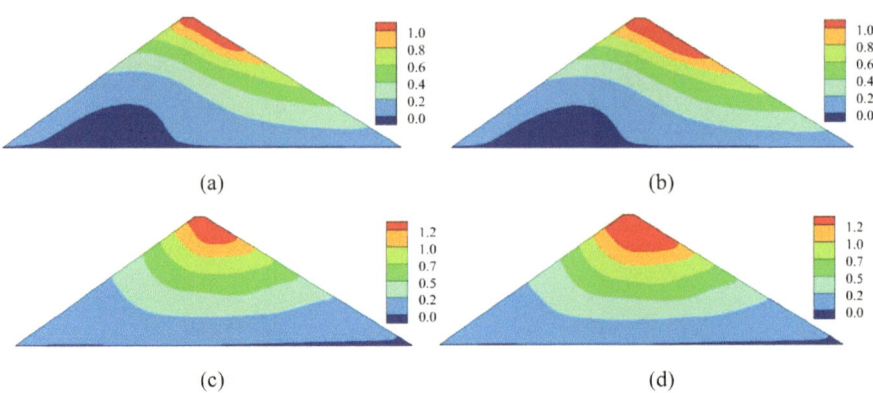

Fig. 4.7 Horizontal and vertical residual deformation, **a** horizontal residual deformation of single sample, **b** mean of horizontal residual deformation, **c** vertical residual deformation of single sample, **d** mean of horizontal residual deformation

of the cloud map and the rule of change in Sect. 3.4, but the dispersion is smaller than that caused by considering the randomness of the ground motion response; when the normal distribution, the horizontal residual deformation of the top of the dam has a minimum value of 0.8 m, a maximum value of 1.6 m, a difference of about 1 times; when considering different distribution types of rockpile parameters, the horizontal residual deformation caused by normal distribution and lognormal distribution does not differ much, but there is a certain difference at the top of the dam, and the mean values are 1.2 and 1.3 m, respectively, with a difference of about 8%. Figure 4.9 shows the distribution law of vertical residual deformation along the dam height. It can be seen that the vertical displacement response along the dam height caused by different material parameters has different distribution laws, but the trend is basically the same, but the dispersion is smaller than the response caused by the randomness of the ground motion; normal distribution, the maximum value of the vertical residual deformation at the top of the dam is 2.0 m, and the minimum value is 0.8 m, which is a difference of about 1.5 times; considering the different types of distribution of the parameters of the stacked rock material, the vertical residual deformation caused by the normal distribution and the log normal distribution is also not much different from that caused by the normal distribution. Vertical residual deformation caused by normal distribution and lognormal distribution when considering different types of distribution of rock pile parameters is also not much difference, the average value of the top of the dam is 1.5m and 1.7m respectively, with a difference of about 13%.

Figure 4.10 shows the horizontal displacement time history of the dam roof and their mean and standard deviation time history, and Fig. 4.11 shows the corresponding time history of the vertical displacement. It can be seen that the deformation response and ground motion randomness show different patterns of change, indicating the necessity of considering the randomness of the material parameters; the drastic change of the standard deviation indicates that under the influence of the ground motion, the randomness of the material parameters has a more drastic effect on the

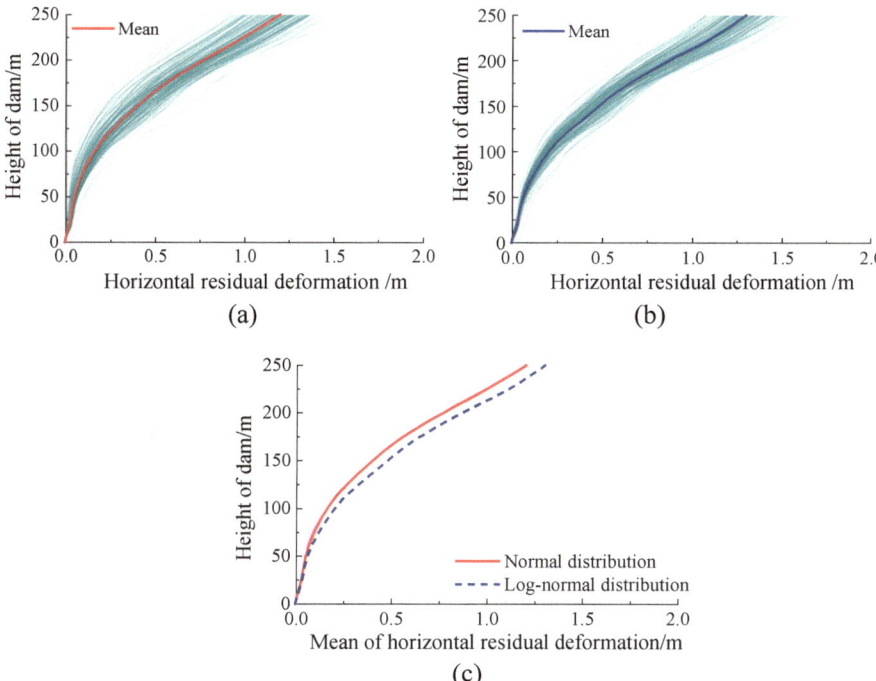

Fig. 4.8 Distribution of horizontal residual deformation along the dam height. **a** Normal distribution of material parameter, **b** log-normal distribution of material parameters, **c** comparison of mean values of horizontal residual deformation due to different distribution types

deformation response; considering the normal distribution of the material parameters and the lognormal distribution of the acceleration response of the time history of the law of the change is basically the same, indicating that the material parameters of the different and different distribution type The different deformation law response basically has no effect on the different material parameters, and only has a small effect on the numerical size.

From the discrete point distribution and exceeding probability of horizontal residual deformation at the dam top shown in Fig. 4.12, it can be seen that the horizontal residual deformation response is different with different material parameters, and there is a certain difference; for normal distribution, the maximum value is 1.6 m, and the minimum value is 0.8 m, which is a difference of about 1 times; considering the 95% exceeding probability and the 5% exceeding probability, there is a difference of about 0.7 times; and for the lognormal distribution, the maximum value is 1.6 m, minimum value 0.9 m, the difference of horizontal residual deformation under the mean value and each exceeding probability in the case of different distribution types is also not big, within 10%; therefore, the randomness of material parameters should be considered for the influence of horizontal displacement response, but the influence of the distribution types of material parameters can be

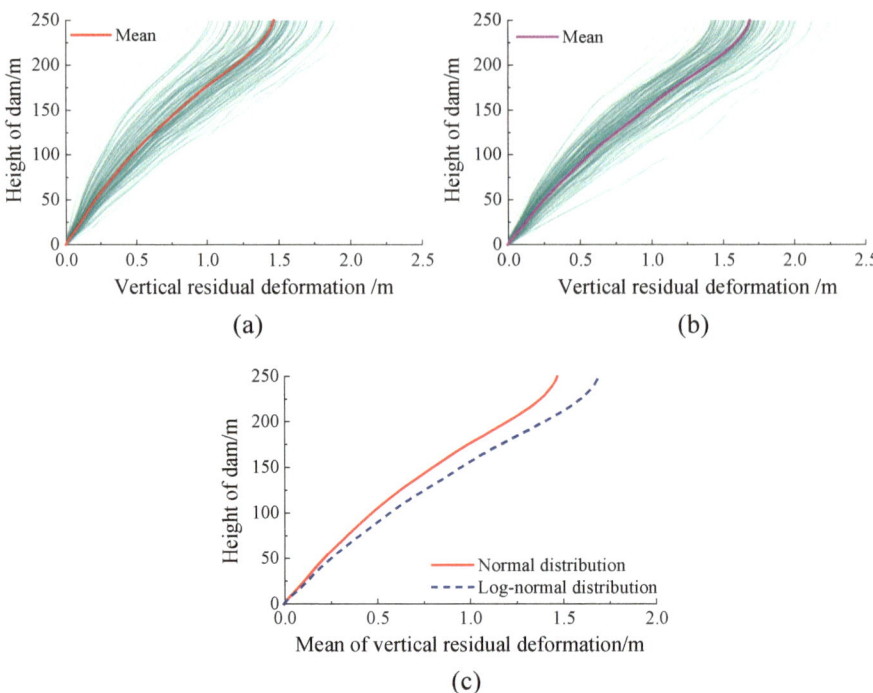

Fig. 4.9 Distribution of vertical residual deformation along the dam height. **a** Normal distribution of material parameter, **b** log-normal distribution of material parameters, **c** comparison of mean values of vertical residual deformation due to different distribution types

disregarded. Figure 4.13 shows the discrete point distribution and beyond the probability plot of the vertical residual deformation of the top of the dam, and it can be seen that the vertical displacement response is different for different material parameters, and the maximum value is 2.0 m and the minimum value is 0.8 m for the normal distribution; the difference is about 1 times when considering the 95% beyond the probability and the 5% beyond the probability; and the maximum value is 2.3 m and the minimum value is 1.1 m for the lognormal distribution and the maximum value is 2.3 m and the minimum value is 1.1 m for the case of different types of distributions. The maximum vertical displacement response value under the mean value and each exceeding probability also has some difference, around 15%; therefore, the randomness of material parameters should be considered for the effect of vertical displacement, and to a certain extent, the effect of the type of distribution of material parameters on it should be considered.

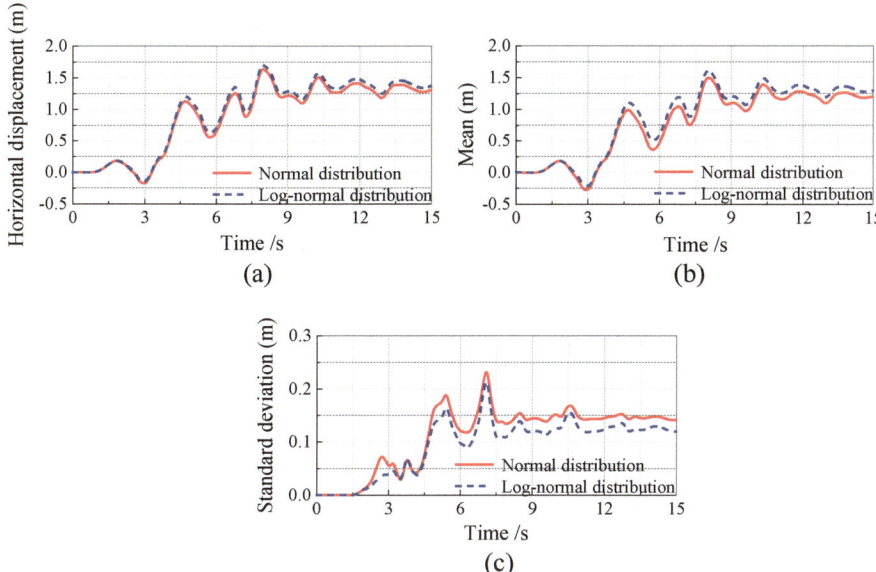

Fig. 4.10 Horizontal displacement time history and their mean and standard deviation, **a** typical sample, **b** mean, **c** standard deviation

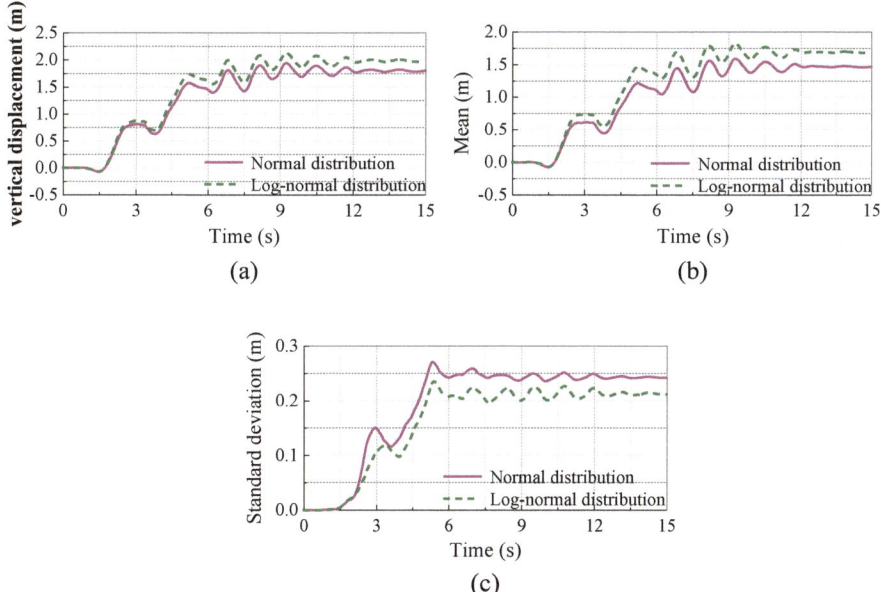

Fig. 4.11 Vertical displacement time history and their mean and standard deviation, **a** typical sample, **b** mean, **c** standard deviation

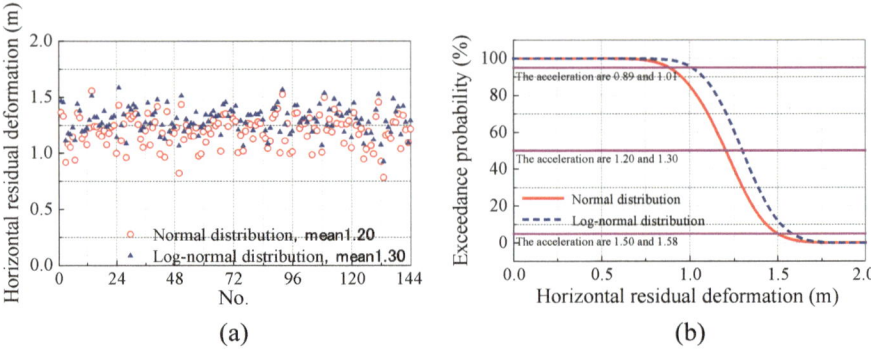

Fig. 4.12 Discrete point distribution of maximum horizontal displacement and exceedance probability, **a** horizontal residual deformation discrete point distribution, **b** horizontal residual deformation exceeding probability

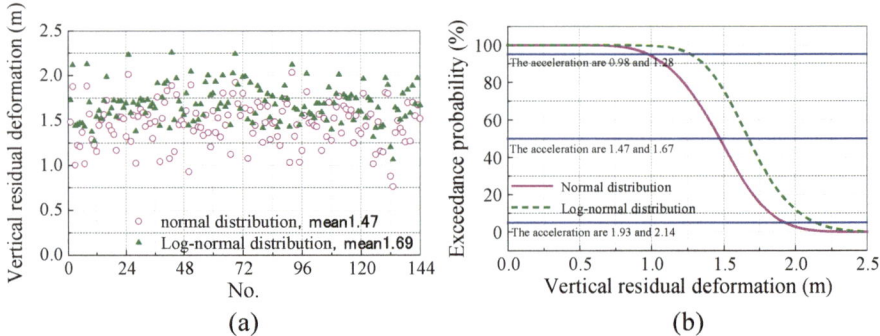

Fig. 4.13 Discrete point distribution of maximum vertical displacement and exceedance probability, **a** vertical residual deformation discrete point distribution, **b** vertical residual deformation discrete point distribution

4.3.4 Face-Slab Stress

Figure 4.14 shows the distribution pattern of the maximum downslope stress of the panel along the dam height (tensile stress is positive, compressive stress is negative). It can be seen that the distribution of stress response along the dam height caused by different material parameters is different, but the trend is basically the same; the average value of downslope tensile stress is mainly concentrated in the range of 100–250 m (0.4–1.0H), and reaches the maximum value around 0.75H, which is similar to that of the ground motion stochasticity; the dispersion of the stress caused by the material parameter stochasticity is smaller than that of the ground motion stochasticity, but the maximum downslope stress at the bottom of the dam is smaller than that of the ground motion stochasticity. Dispersion is small, but the maximum downslope stress at the bottom of the dam is more discrete; the location and value of the maximum tensile stress are different, so it is necessary to analyze the panel stress law from the perspective of the material parameter uncertainty, but the type of distribution of the material parameter does not have much effect on the stress response. From the distribution of stress rms along the dam height at 3, 9 and 15 s moments (Fig. 4.14d), the panels are mainly subjected to compressive stresses when analyzed from the rms point of view and that the material parameter normal and lognormal distributions do not have a significant effect on the stress response.

A comparison of the time histories of variation of the mean and standard deviation of the stress response considering normal and lognormal distributions of the material parameters, given in Fig. 4.15, shows that the type of material distribution has little effect on the stress response.

Figure 4.16 shows the exceeding probability of cumulative overstress holding time (COD) for different demand stress ratios (DCR) and the performance states obtained based on this. The cumulative overstress holding time due to different distribution types does not differ much, and the damage states and probabilities are closer to each other, with almost complete damage under 0.5g ground motion.

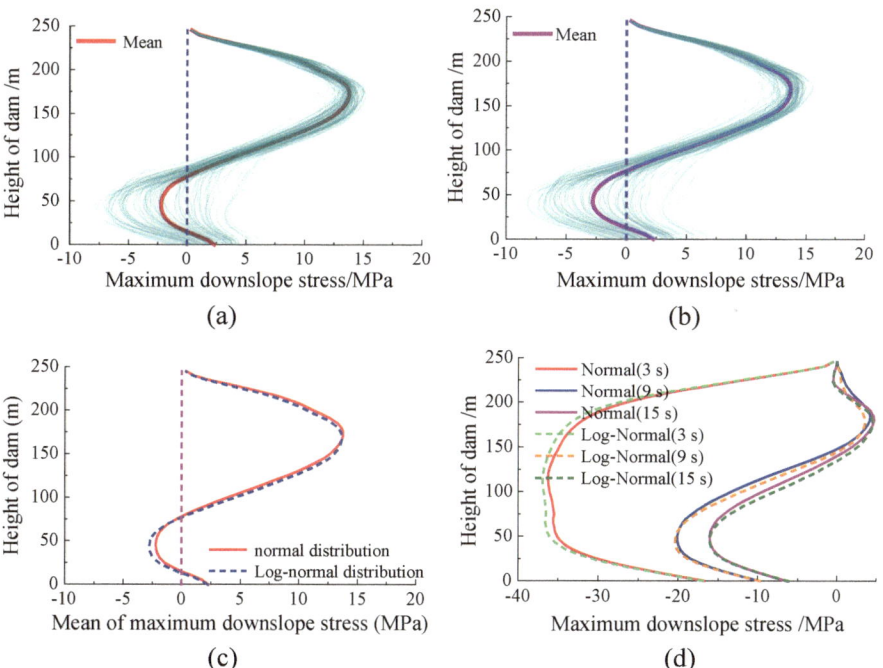

Fig. 4.14 The maximum stress distribution along the dam height, **a** normal distribution, **b** lognormal distribution, **c** comparison of mean values of downslope stress induced by different distribution types, **d** mean value of downslope stress at different moments

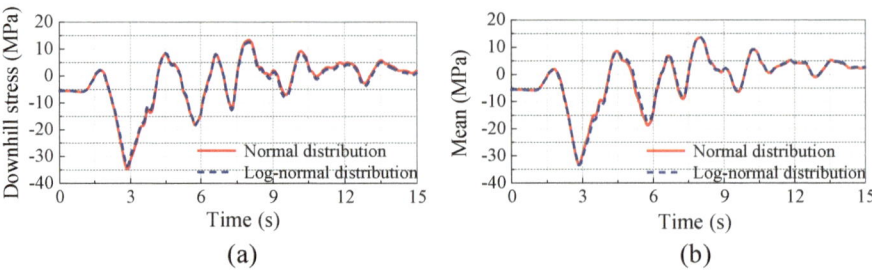

Fig. 4.15 Stress response, mean and standard deviation time history, **a** mean, **b** standard deviation

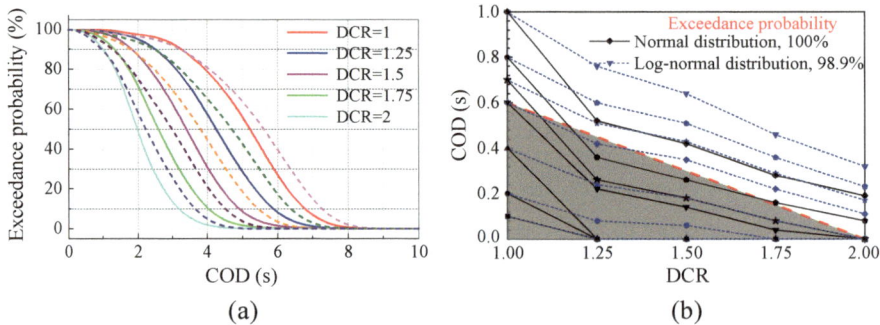

Fig. 4.16 Exceedance probability of faced-slab damage index, **a** exceedance probability, **b** damage level evaluation criteria

4.3.5 Performance-Based Seismic Safety Evaluation

Through the above probability analysis, the damage probability of each performance index under PGA $= 0.5$g seismic intensity is obtained. From the point of view of dam deformation, according to the criteria of damage classification in Chap. 3 (relative seismic subsidence rate of the dam top 0.3%, mild damage; 0.7%, moderate damage; 1.0%, severe damage), the corresponding damage probabilities are 100% (normal distribution), 100% (lognormal distribution), 15.8% (normal distribution), 38.4% (lognormal distribution), 0% (normal distribution), and 0% (lognormal distribution). The demand stress ratio from the panel combined with the cumulative overstress holds when it is essentially destroyed.

4.4 Conclusion

In this chapter, considering the uncertainty of the material parameters of rockpile materials, the high-dimensional elastic–plastic stochastic parameter samples are generated by the GF-deviation reforming technique, and jointly with the generalized probability density evolution method, the effects of the parameter randomness and its different types of distributions on the responses of acceleration, deformation, and panel stresses of high panel rockpile dams are investigated in detail from the stochastic dynamics and probabilistic points of view. The main work and conclusions are as follows:

According to the results of stochastic dynamic analysis in this chapter, under the action of deterministic ground motion (PGA = 0.5g), analyzed from the perspective of stochastic dynamics and probability, the uncertainty of material parameters has a large impact on the acceleration, deformation and panel stress response of CFRD, and the response values are mainly varied in this interval by taking the 5% exceeding probability and the 95% exceeding probability as a criterion, and the maximum value is about 1.5–2 times of the minimum value The maximum value is about 1.5–5 times of the minimum value, which is slightly less discrete than the seismic response caused by the randomness of the ground motion; however, the comparison of the response values by considering the normal distribution of the material parameters and the lognormal distribution shows that the distribution type does not have much influence on the acceleration and stress response, and the difference is basically within 10%, but has a slightly larger influence on the deformation response. Therefore, under deterministic ground motion excitation, it is necessary to consider the randomness of material parameters, but only the effect of distribution type needs to be properly considered. In future studies, it is necessary to determine the distribution types of the material parameters of the rock piles and explore the correlations between them through many tests and sample statistics, and then to perform stochastic dynamic and probabilistic analyses.

Chapter 5
Stochastic Dynamic Analysis of CFRD Considering the Coupling Randomness of Ground Motion and Material Parameters

5.1 Introduction

Currently, in probabilistic fragility analysis, the randomness of ground motion and material parameters is typically addressed by randomly combining several ground motions selected from seismic databases and sampled material parameters (Xu et al. 2018). However, this approach lacks coupling effects, suffers from insufficient sample sizes or involves extensive computational efforts, thereby limiting the acquisition of comprehensive probability information. There are few studies on the coupling randomness of ground motion and material parameters on the dynamic response of structures, and even less on the seismic response of high CFRDs.

This chapter considers the coupled randomness of ground motion and material parameters by simultaneously generating stochastic ground motion acceleration time histories and random material parameters. Combined with methods such as the generalized probability density evolution method (GPDEM), reliability probability analysis, and fragility analysis, it reveals the impact of coupled randomness on the seismic dynamic response of high CFRDs from the perspectives of stochastic dynamics and probability. Subsequently, it enhances the performance-based seismic safety evaluation framework.

5.2 Basic Information

Considering the coupled randomness of ground motion and material parameters, there are 10 random parameters, including 2 uniformly distributed random variables Θ_1 and Θ_2 in the ground motion spectrum expression in Sect. 2.5 (interval $[0, 2\pi]$ uniformly distributed and independent of each other) and 8 random variables of material parameters in Sect. 4.3.1 (normally distributed and independent of each other, coefficient of variation 0.1). Based on GF-discrepancy method, 144 groups of

© The Author(s) 2025
B. Xu and R. Pang, *Stochastic Dynamic Response Analysis and Performance-Based Seismic Safety Evaluation for High Concrete Faced Rockfill Dams*, Hydroscience and Engineering, https://doi.org/10.1007/978-981-97-7198-1_5

stochastic ground motion and random material parameter samples are simultaneously generated. The vertical seismic acceleration is taken as 2/3 of the horizontal seismic acceleration, and the peak acceleration of bedrock along the river is adjusted to 0.1 g-1.0 g, respectively, with an interval of 0.1 g. A series of finite element dynamic calculations were carried out with the input of ground motion. A total of 1440 working conditions were calculated for 10 peak accelerations of ground motion to obtain the stochastic dynamic information under the action of different seismic intensities. Then the probability density evolution equation was solved based on the above numerical method to obtain the probability information of seismic response of high CFRDs at each time. In the following, the effects of the coupling randomness of ground motion and material parameters are analyzed in detail from the aspects of dam acceleration, deformation and random dynamic response of the slope stress of the panel, as well as the probability of these physical indicators, and the corresponding performance-based seismic safety evaluation framework is established. The finite element model, loading conditions, and other material parameter information in this chapter are completely consistent with Sect. 3.4.1.

5.3 Stochastic Dynamic and Probabilistic Analysis of High CFRD

5.3.1 Dam Acceleration

Figure 5.1 shows the maximum horizontal acceleration response based on a single sample and the mean distribution of 144 groups of sample responses at PGA = 0.5 g, respectively. It can be seen that the distribution law is similar to that considering only the randomness of ground motion, but it is different from that considering only the randomness of parameters, indicating that the randomness of ground motion plays a more important role in acceleration response.

In order to further study the influence of the coupled randomness of ground motion and material parameters on the acceleration response of the dam, Fig. 5.2 illustrates the distribution curve and average value of the maximum horizontal acceleration response with the dam height under the seismic intensity of 0.5 g. It can be seen that under the action of coupled randomness, the distribution of acceleration is similar

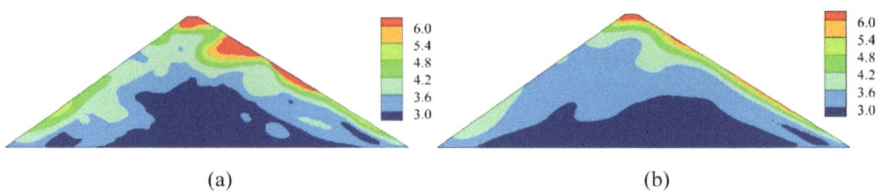

(a)   (b)

Fig. 5.1 Maximum horizontal acceleration response. **a** A single sample. **b** Mean value

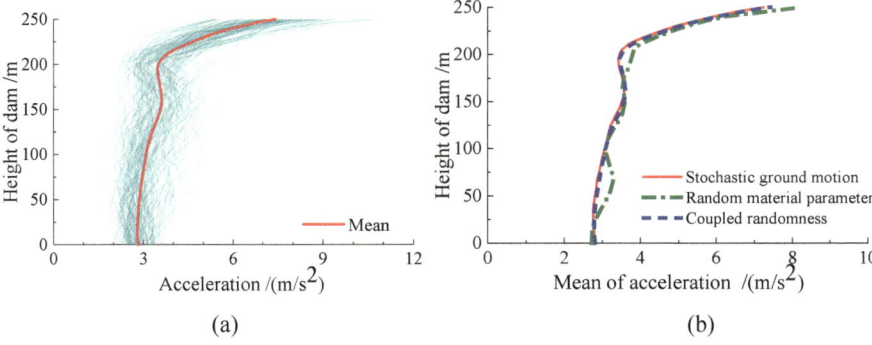

Fig. 5.2 Distribution of maximum horizontal acceleration along dam height. **a** Response to coupled randomness. **b** Comparison of responses under different randomization factors

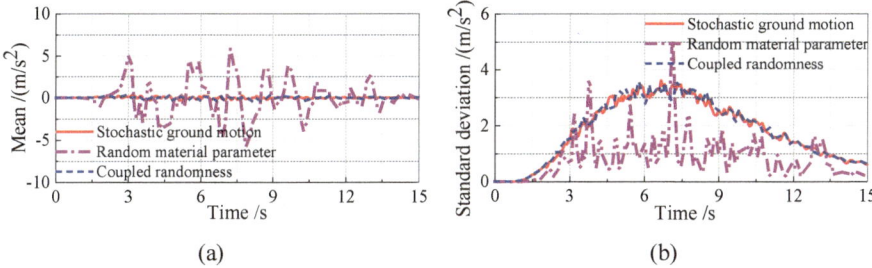

Fig. 5.3 Mean and standard deviation time-history of horizontal acceleration of dam crest based on different randomness. **a** Mean. **b** Standard deviation

to that of ground motion, which is more obvious in the mean value. From the mean value, the distribution of acceleration response along dam height caused by coupled randomness basically coincides with the response caused by ground motion randomness. The means and standard deviation time history in Fig. 5.3 again proves the above conclusion, but it is quite different from the response considering only the random material parameters.

5.3.2   Dam Deformation

Figure 5.5 depicts the residual horizontal deformation responses based on single-group samples and the averaged responses from 144 sample groups under PGA = 0.5 g. Figure 5.6 illustrates the vertical residual deformation. It is evident that regardless of the response from single-group samples or the averaged response from multiple sample groups, both horizontal and vertical residual deformations occur

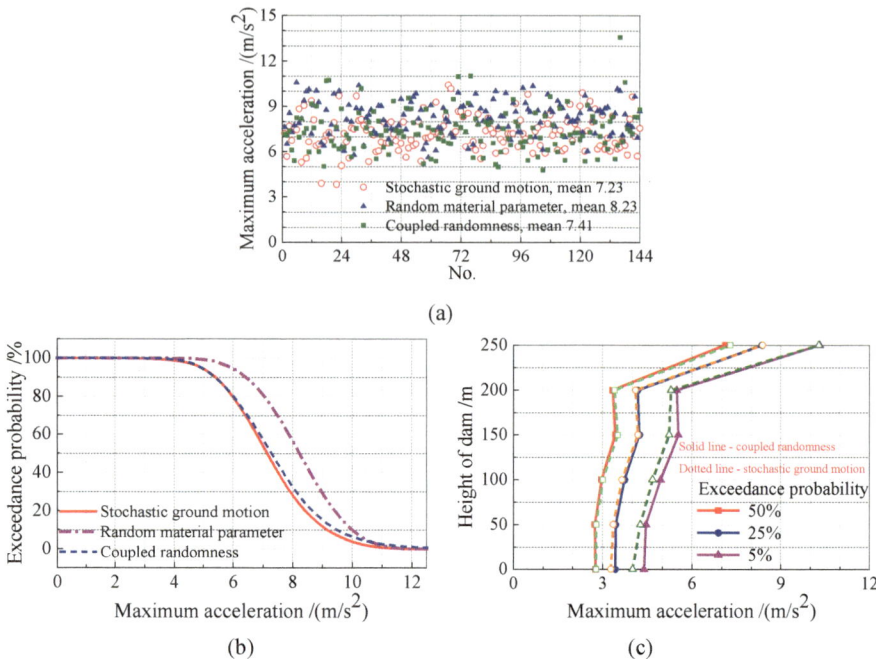

Fig. 5.4 Discrete point distribution, exceedance probability of maximum acceleration. **a** Scatter distribution of maximum acceleration. **b** Exceedance probability of maximum acceleration. **c** Distribution of maximum acceleration along the dam height under typical exceedance probabilities

at the crest of the dam. Furthermore, these deformations closely align with the distribution pattern of responses based on seismic randomness.

Figure 5.7 displays the distribution pattern of horizontal displacements along the dam height at a seismic intensity of 0.5 g at the final moment. It can be observed that the distribution pattern based on the coupling of seismic and material parameters randomness is similar to that considering seismic randomness. This similarity is particularly evident in the mean distribution along the dam height, as the two lines nearly coincide. However, there is a notable difference when compared to the distribution pattern considering only material parameter randomness. Figure 5.8 illustrates

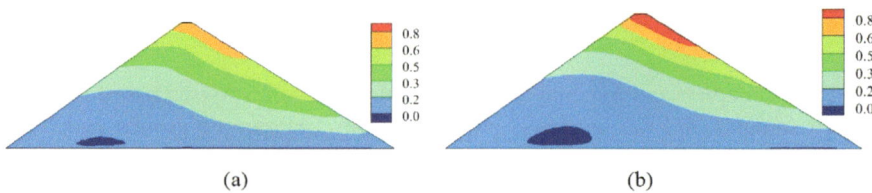

Fig. 5.5 Horizontal residual deformation. **a** Horizontal residual deformation of a single group of samples. **b** Mean horizontal residual deformation

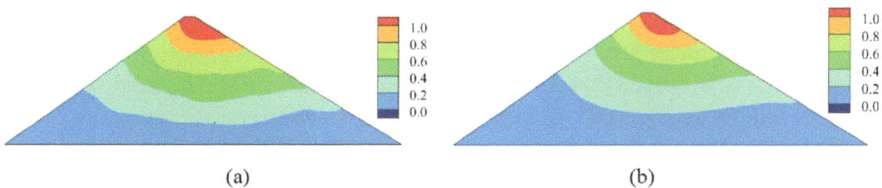

Fig. 5.6 Vertical residual deformation. **a** Vertical residual deformation of one group of samples. **b** Mean vertical residual deformation

the distribution pattern of vertical displacements along the dam height at a seismic intensity of 0.5 g at the final moment. Similarly, it can be concluded that seismic randomness seems to exert a controlling influence on the vertical deformation, but there is a slight difference in the numerical values at the crest. The vertical deformation based on coupled randomness is 1.16 m, while the vertical deformation based on seismic randomness is 1.20 m, indicating a difference of approximately 3.4%.

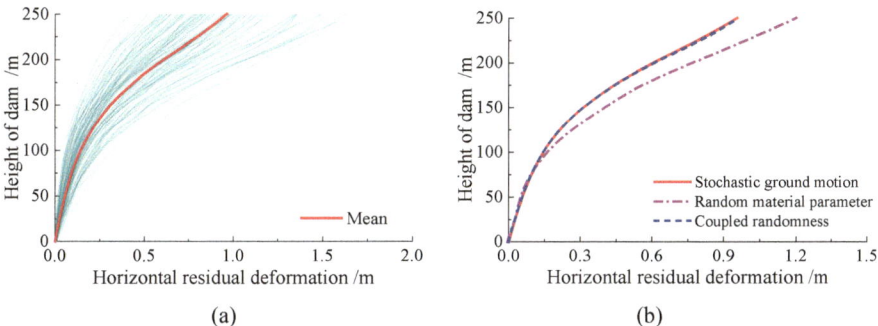

Fig. 5.7 Distribution of horizontal residual deformation along the dam height. **a** Coupled randomness. **b** Comparison of responses under different random factors

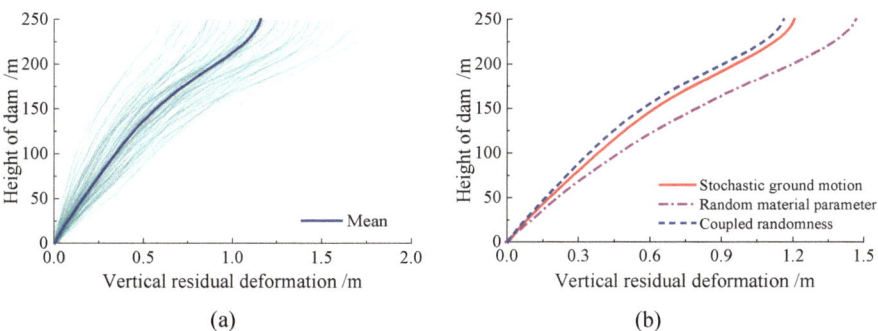

Fig. 5.8 Distribution of vertical residual deformation along the dam height. **a** Coupled randomness. **b** Three random factors

From the horizontal displacement in Fig. 5.9 and the vertical displacement in Fig. 5.10, both mean and standard deviation time histories are presented (PGA = 0.5 g). It can be observed that during the seismic process, the horizontal displacement exhibits almost no difference compared to the seismic response induced by seismic randomness. The temporal pattern of vertical displacement is largely consistent, differing mainly in numerical values. However, both of these responses significantly deviate from the randomness attributed to material parameter randomness. Upon observing the standard deviation time history, it becomes apparent that the standard deviation time history of seismic responses due to the coupling of seismic input and material parameter randomness is notably similar to that induced by seismic input randomness. Nevertheless, there exists a certain discrepancy in terms of numerical values.

The scatter plots in Fig. 5.11, illustrating the residual horizontal and vertical deformations at the crest (PGA = 0.5 g), further corroborate the aforementioned observations. However, it is notable that the deformation responses prompted by seismic randomness and coupled randomness exhibit greater variability compared to the deformations induced by material parameter randomness. This suggests that seismic randomness exerts a more significant influence on dam deformation responses, indicating a certain controlling effect.

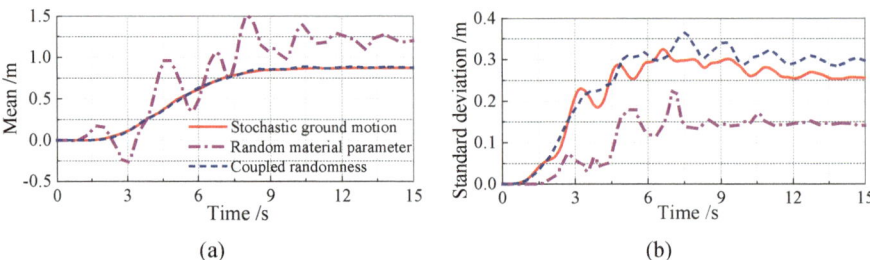

Fig. 5.9 Mean and standard deviation time history of horizontal displacement. **a** Mean time history. **b** Standard deviation time history

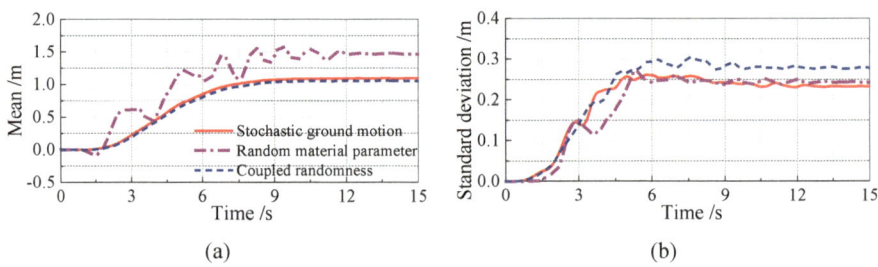

Fig. 5.10 Mean and standard deviation time history of vertical displacement. **a** Mean time history. **b** Standard deviation time history

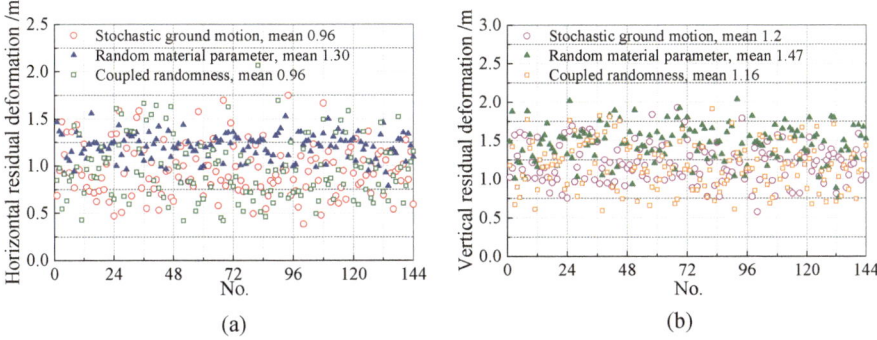

Fig. 5.11 Discrete point distribution of horizontal and vertical deformation. **a** Horizontal residual deformation. **b** Vertical residual deformation

Figure 5.12 presents the exceedance probability curves of residual horizontal and vertical deformations for various seismic intensities (PGA = 0.1–1.0 g, in increments of 0.1 g). It is evident that the exceedance probability curves for residual deformations induced by the coupling of seismic input and material parameters randomness closely align with those resulting from seismic randomness alone. However, as seismic intensity increases, the difference in exceedance probability based on horizontal displacement slightly grows, while the difference based on vertical displacement diminishes slightly.

From the distribution pattern of maximum downslope stress along the dam height under a seismic intensity of 0.5 g in Fig. 5.13 and the variation of downslope stress mean over time in Fig. 5.14, it can be observed that there is minimal difference between the stress responses induced by the coupling randomness and those caused by seismic randomness. This indicates that seismic input plays a primary role in faced-slab stress response. Figure 5.15, which illustrates the exceedance probability of cumulative overstress duration (COD) for various Demand-Capacity Ratios (DCRs) and the resultant performance states, further validates this point. Hence, the

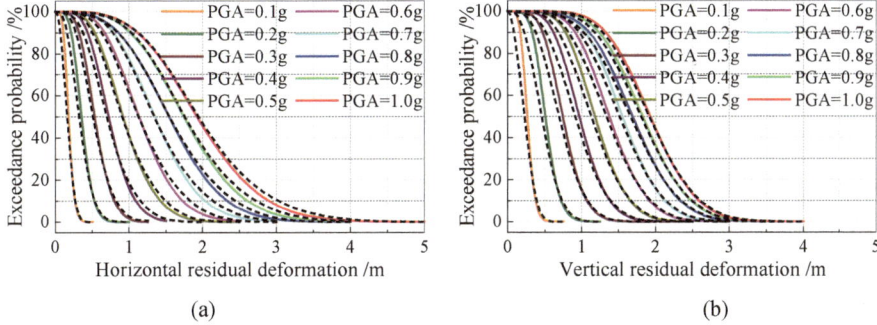

Fig. 5.12 Exceedance probability of horizontal and vertical residual deformation Faced-slab Stress. **a** Horizontal residual deformation. **b** Vertical residual deformation

Fig. 5.13 The maximum downslope stress distribution along the dam height

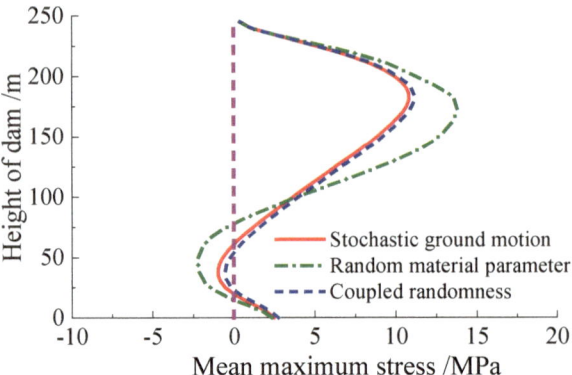

analysis of faced-slab failure probability can focus solely on the impact of seismic randomness.

Fig. 5.14 The mean downslope stress time history

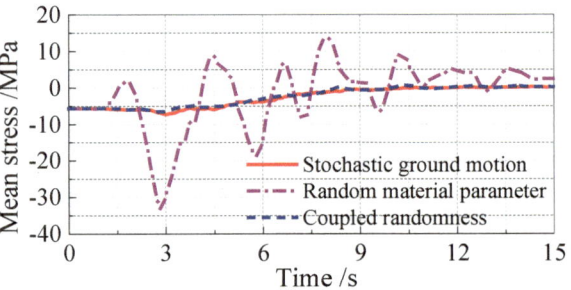

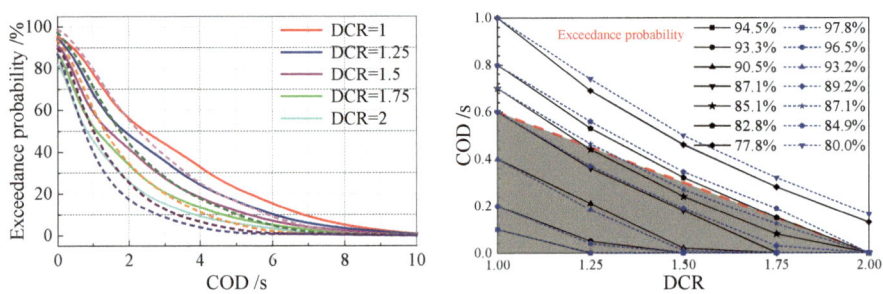

Fig. 5.15 Exceedance probability of faced-slab damage index

5.4 Conclusion

This chapter investigates in detail the effects of coupled randomness of ground motion and material parameters on the dynamic response and seismic safety of high CFRDs from the perspectives of stochastic dynamics and probability. Firstly, stochastic ground motion time histories and random material parameter samples are generated simultaneously by combining spectral representation-random function and material parameter random variables. Then, the stochastic dynamic response and probability characteristics of high CFRDs under the influence of coupled randomness of ground motion and material parameters are analyzed. By comparing the stochastic dynamic and probability results of dam body acceleration, deformation, and panel stress caused by seismic randomness, material parameter uncertainty, and coupled randomness, it is demonstrated that the seismic randomness has a predominant control effect on seismic response, and the influence of material parameter randomness can be neglected to some extent when conducting comprehensive random dynamic response analysis of dams. Finally, based on the above performance indicators and performance levels, a seismic safety evaluation framework is established for different seismic intensity levels, multiple seismic intensity-multiple performance objective-surpassing probability scenarios. Vulnerability curves for different damage levels are obtained. The results show that there is little difference in performance safety evaluation probability obtained from seismic randomness and coupled randomness, within 10%. Therefore, when performing performance-based seismic safety evaluation of high CFRDs, considering seismic randomness alone is sufficient to meet the requirements.

The above chapter mainly conducts seismic safety analysis of two-dimensional high CFRDs from the perspectives of stochastic dynamics and probability. However, three-dimensional effects have a certain influence on the seismic response of high CFRDs, especially for relatively thin panels. Two-dimensional analysis often struggles to accurately describe and define their failure states. Nevertheless, the above research can qualitatively and quantitatively study and compare panel failure states to some extent from the perspectives of stochastic dynamics and probability. Additionally, three-dimensional analysis is commonly used in engineering applications to study the deformation of dam bodies, especially the failure patterns of panels. Therefore, the following research will explore the stochastic dynamic response and probability variation laws of high CFRDs based on three-dimensional numerical calculations, revealing the relationship with the aforementioned two-dimensional stochastic dynamic analysis, and thereby establishing corresponding methods and frameworks for performance safety evaluation.

Reference

Xu B, Zhang X, Pang R et al (2018) Seismic performance analysis of high core-wall rockfill dams based on deformation and stability. J Hydroelectric Eng 37(10):31–38

Chapter 6
Stochastic Seismic Response and Performance Safety Evaluation for 3-D High CFRD

6.1 Introduction

It is widely recognized that high CFRDs, especially those with thinner slabs, exhibit significant three-dimensional effects. Three-dimensional finite element analysis offers a more realistic depiction of the stress distribution of the dam. However, there is a lack of studies that utilize stochastic dynamics and probabilistic analyses to investigate the seismic safety of three-dimensional CFRDs and establish corresponding safety evaluation criteria. Building upon the analyses conducted in the preceding chapters, it is evident that the randomness of ground motions is the primary factor influencing the stochastic dynamic response of CFRDs. In this chapter, drawing from the performance-based seismic safety evaluation studies of two-dimensional high CFRDs presented in Chaps. 3 and 5, we delve into the distribution and variations of acceleration, deformation, and slab stress of three-dimensional high CFRDs, considering the stochastic dynamics and probability perspectives. Additionally, we elucidate the interconnection with two-dimensional analyses. Ultimately, we identify dam crest subsidence and the safety of face-slabs impermeable body as key indicators for defining the limit states of CFRDs. Moreover, we enhance the performance-based framework for seismic safety evaluation to assess the reliability of CFRDs.

© The Author(s) 2025
B. Xu and R. Pang, *Stochastic Dynamic Response Analysis and Performance-Based Seismic Safety Evaluation for High Concrete Faced Rockfill Dams*, Hydroscience and Engineering, https://doi.org/10.1007/978-981-97-7198-1_6

6.2 Basic Information of High CFRDs

The 250 m three-dimensional CFRD finite element mesh model is shown in Fig. 6.1. The upstream and downstream slopes have a ratio of 1:1.4 and 1:1.6. The width of dam crest is 18 m, the thickness of concrete face-slabs is 1.175 m. The horizontal width of the cushion and transition areas setting under the concrete slab is 4 and 6 m, respectively. The slab was poured in three phases, 75 m, 150 m and 250 m respectively. The dam was filled in 50 layers, and the water storage was divided into 48 steps up to 240 m. The element simulating the dam comprises hexahedral iso-parametric elements and a minor proportion of degenerate tetrahedral elements. The Goodman contact surface unit without thickness is set up between face-slabs and cushion Goodman (1968). Moreover, vertical and peripheral joints are simulated using 8-node spatial jointing element. Seismic input using fluctuation input method based on viscoelastic artificial boundary setting. To simplify the calculation, bedrock is not considered in this study. The additional mass method in Chap. 3 is used to analyze the hydrodynamic pressure on slabs. Both static and dynamic processes are simulated using a uniform generalized plastic model for rockfill, transition, and cushion materials. The contact surface between face-slabs and cushion can be simulated using the generalized plastic contact surface model. Material parameters using values from Sect. 3.4.1. The face-slabs simulated by a linear elastic model for C30 concrete. The following parameters were used: $E = 3.1 \times 10^4$ MPa, $\rho = 2.40$ g/cm3, $\nu = 0.167, f_c = 27.6$ MPa.

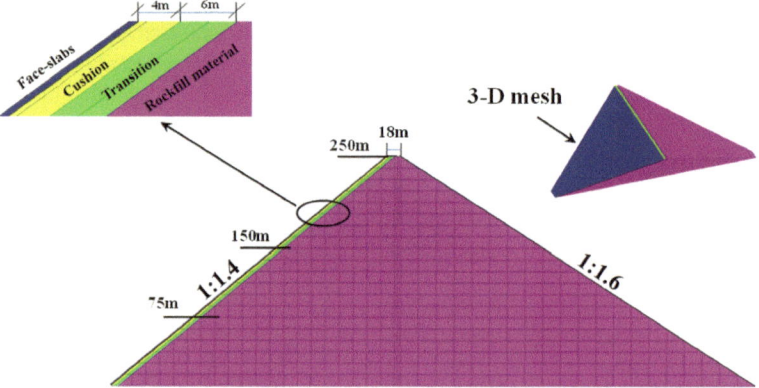

Fig. 6.1 Finite element mesh model of the high CFRD

6.3 Stochastic Dynamic Response and Probability Analysis of High CFRDs

Non-stationary random ground motions were simulated using the intensity-frequency fully non-stationary stochastic ground motion generation methods. The ground motions were generated at PGA intervals of 0.1 g, ranging from 0.1 g to 1.0 g. For each PGA, 144 ground motions were created, resulting in a total of 1440. In addition, the vertical acceleration is two thirds that of the horizontal acceleration. The stochastic dynamic response and the associated probability information under different intensity ground motions is obtained by performing finite element calculations and probabilistic density evolution analysis on the model.

6.3.1 Dam Acceleration

Figure 6.2 shows the distribution of the maximum acceleration of the dam based on a single sample of ground motion at PGA of 0.5 g. Figure 6.3 shows the distribution of the mean values of the 144 acceleration responses. Regardless of the individual response or the mean value, the amplification of acceleration response is most obvious in the dam crest region. In addition, the downstream dam slope is also the concentration area of larger acceleration response. This may be due to the confinement of the upstream slab and the reflection of the fluctuation of the downstream dam slope, essentially the same as the two-dimensional analysis. From the mean value analysis, the dam acceleration amplification is circularly or elliptically distributed in the top region of the dam and roughly symmetrically distributed along the dam axis with respect to the center line of the river valley. The amplification effect is most obvious at the dam crest location at the river valley. It also indicates that the stochastic ground excitation has a great influence on the acceleration response of the dam body.

Figure 6.4 illustrates distribution of the maximum acceleration and amplification in downstream direction along dam height under different seismic intensities, respectively. Different ground motions cause different response curves, but the curves have the same trend. A sudden alteration happens at around 0.8H, the amplification effect at the top of the dam is particularly pronounced, showing a strong "whip-sheath" effect. The acceleration amplification under 0.5 g PGA in Fig. 6.4c also shows a large dispersion. These features are similar to the 2-D acceleration response. Meanwhile, Fig. 6.4d shows that the amplification of the mean acceleration tends to decrease as the intensity of ground motion increases. Therefore, it is necessary to investigate the acceleration response of CFRDs based on stochastic ground motions.

The horizontal coordinates of Fig. 6.5 are the normalized dam crest axis and the vertical coordinates are the down-river acceleration at the dam crest. Figure 6.5d shows the amplification of the mean acceleration corresponding to earthquakes of different intensities. The acceleration amplification effect is not obvious near banks, but increases suddenly near the middle of the valley, showing obvious

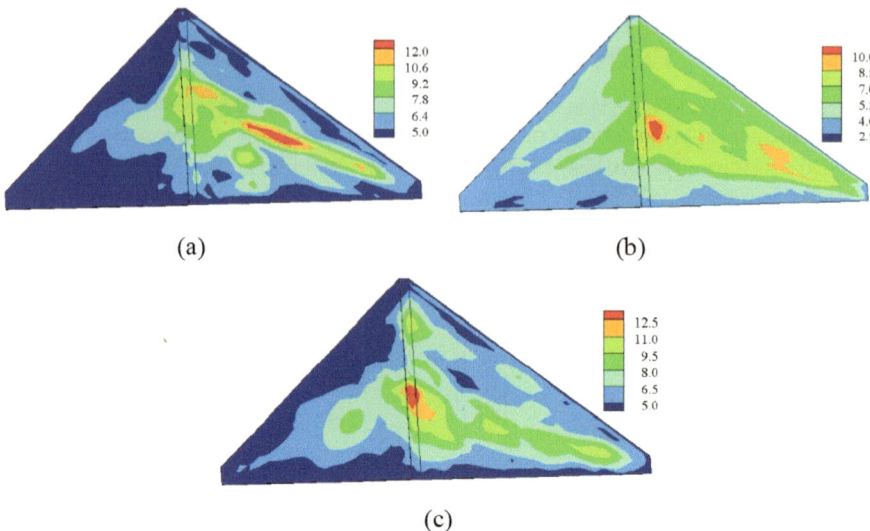

Fig. 6.2 Dam acceleration distribution based on the single sample. **a** Horizontal direction **b** Vertical direction **c** Dam axial direction

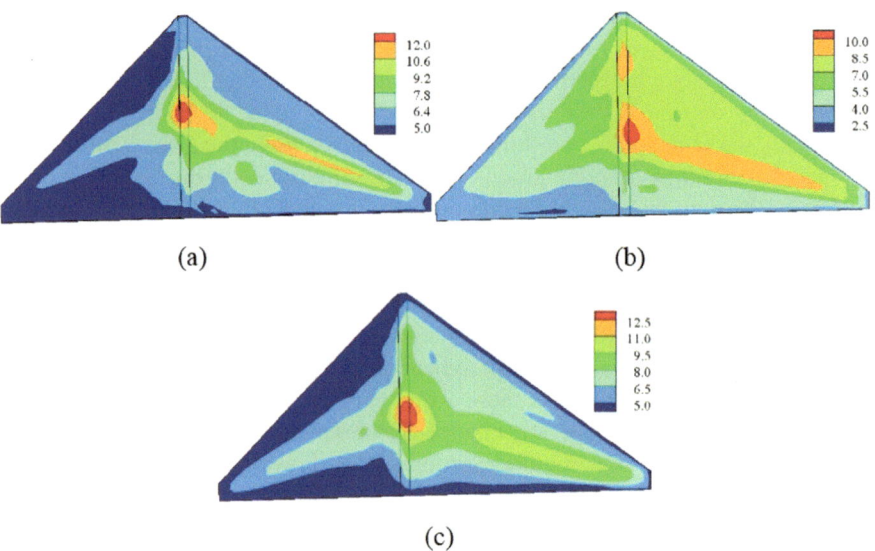

Fig. 6.3 Dam acceleration mean distribution based on the 144 samples. **a** Horizontal direction **b** Vertical direction **c** Dam axial direction

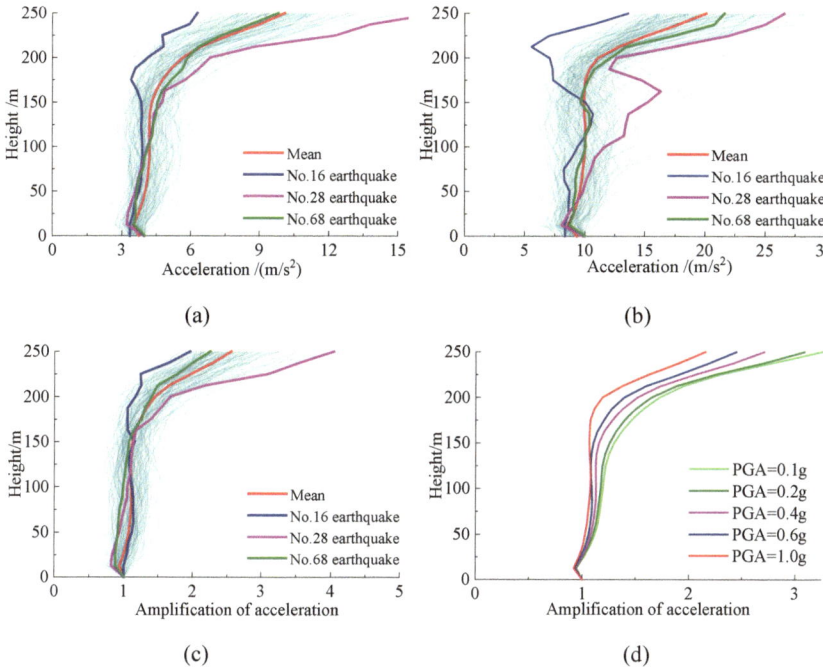

Fig. 6.4 Distribution of maximum horizontal acceleration along dam height. **a** PGA = 0.4 g **b** PGA = 1.0 g. **c** PGA = 0.5 g **d** Mean of different PGA

three-dimensional effect, and the response laws of different ground motions are quite different. For the same ground motion, there are certain differences in the response patterns at different intensities. In addition, as the intensity of ground motion increases, the amplification of acceleration along the axial direction of the dam tends to decrease.

Figure 6.6 shows the discrete point of the maximum horizontal acceleration at the top under different seismic intensities, which can provide a reference for the seismic safety design and control criteria for CFRDs. The maximum acceleration distribution is relatively discrete, with 95 and 5% exceedance probabilities. The difference between the maximum and minimum values is large, with reaching 2–2.5 times, which suggests that the acceleration response of the dam body under ground motions generated in this research is statistically significant. The maximum acceleration response under different seismic intensities is basically linearly distributed, similar to the 2-D trend.

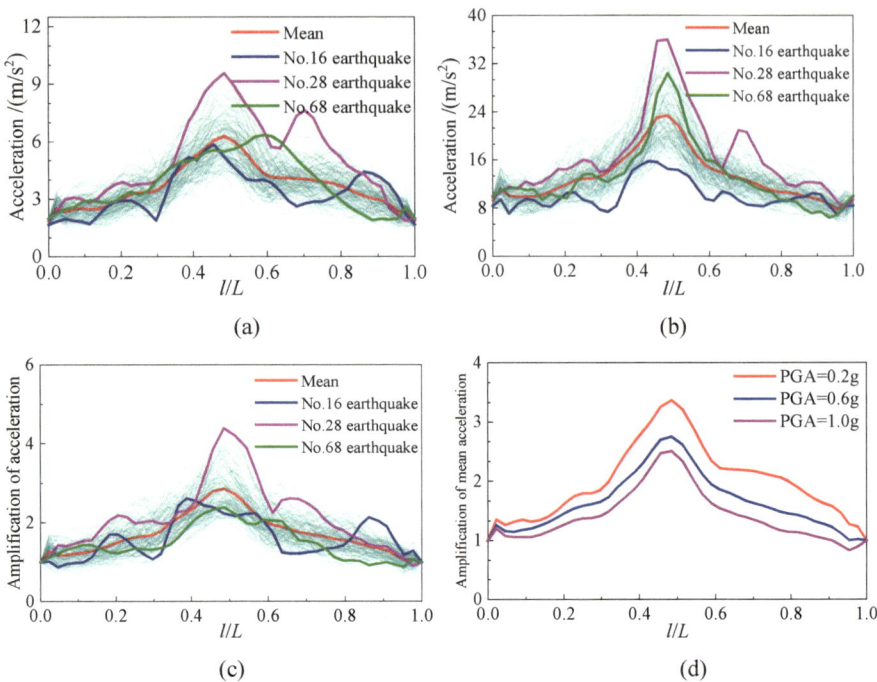

Fig. 6.5 Distribution of maximum horizontal acceleration along dam axial direction. **a** PGA = 0.2 g **b** PGA = 1.0 g **c** PGA = 0.5 g **d** Mean of different PGA

Fig. 6.6 Discrete point distribution of maximum horizontal acceleration

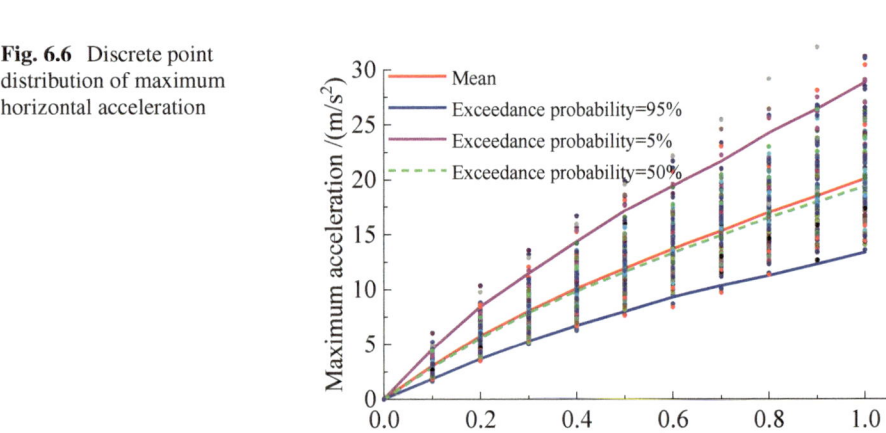

6.3.2 Dam Deformation

Figure 6.7 shows the horizontal and vertical residual deformation responses obtained based on a single sample of ground motion under PGA = 0.5 g, respectively, and

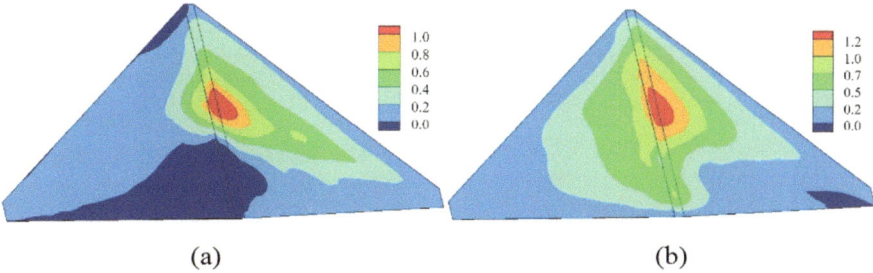

Fig. 6.7 Residual deformation based on the single sample. **a** Horizontal direction **b** Vertical direction

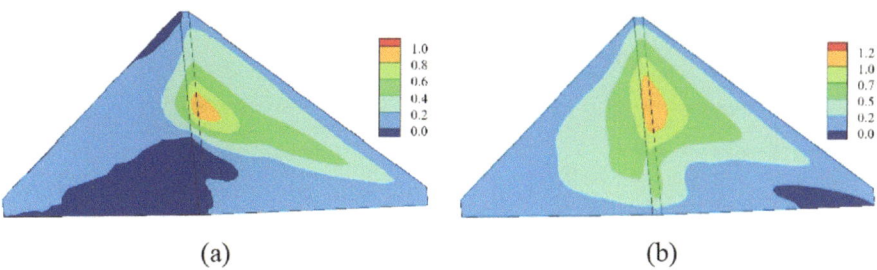

Fig. 6.8 Residual deformation mean based on the 144 samples. **a** Horizontal direction (**b**) Vertical direction

Fig. 6.8 shows the mean of 144 responses. The response occurs mainly in the center of the river valley at the top of the dam, with an approximately elliptical distribution. The result shows a similar pattern to the acceleration responses, indicating a correlation between them.

The exceedance probability curves of horizontal and vertical residual deformation are shown in Fig. 6.9, which serves as a useful point of reference for subsequent evaluations of seismic safety based on deformation. Figure 6.10 shows the distribution of discrete points of residual deformation in the horizontal and vertical direction for different seismic intensities, which has a certain degree of dispersion and is similar to the pattern of the two-dimensional. Table 6.1 lists the horizontal and vertical residual deformations corresponding to different exceedance probabilities under different PGA. Based on the 95% and 5% exceedance probabilities, the maximum value in the horizontal direction is about 3–4 times of the minimum value, and the maximum value in the vertical direction is approximately twice that of the minimum., which can provide reference for CFRDs ultimate seismic capacity analysis under different PGA.

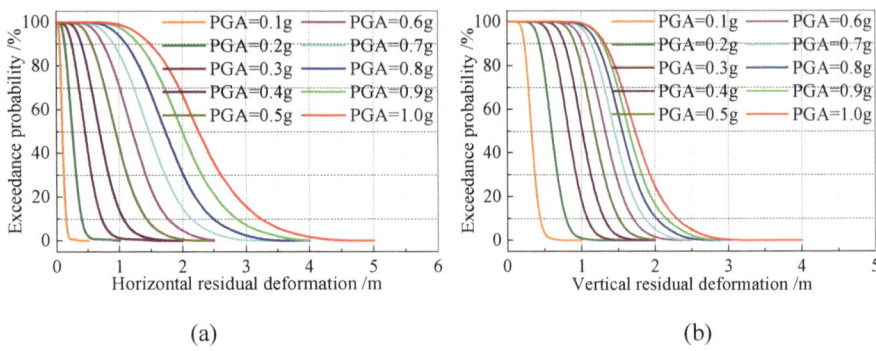

Fig. 6.9 Exceedance probability of horizontal and vertical residual deformation. **a** Horizontal direction. **b** Vertical direction

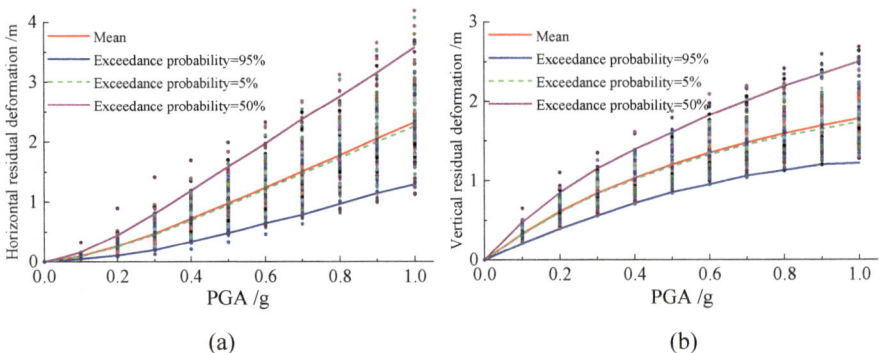

Fig. 6.10 Horizontal and vertical residual deformation under different PGA. **a** Horizontal direction. **b** vertical direction

Table 6.1 The horizontal and vertical residual deformation of dam crest based on different exceedance probability under different PGA

Exceedance probability	PGA (g)										
	0.1	0.2	0.3	0.4	0.5	0.6	0.7	0.8	0.9	1.0	
Horizontal residual deformation (m)	Mean	0.096	0.261	0.477	0.721	0.979	1.241	1.508	1.776	2.052	2.326
	5%	0.160	0.442	0.809	1.187	1.585	1.974	2.379	2.760	3.157	3.578
	50%	0.091	0.249	0.456	0.695	0.948	1.208	1.468	1.728	1.997	2.261
	95%	0.044	0.106	0.196	0.335	0.480	0.644	0.786	0.967	1.144	1.291
Vertical residual deformation (m)	Mean	0.334	0.610	0.840	1.031	1.199	1.347	1.477	1.591	1.688	1.776
	5%	0.477	0.849	1.152	1.392	1.606	1.820	2.005	2.188	2.341	2.495
	50%	0.327	0.600	0.829	1.015	1.179	1.323	1.447	1.556	1.647	1.731
	95%	0209	0.395	0.558	0.715	0.853	0.952	1.059	1.126	1.201	1.218

6.3.3 Overstress Volume Ratio and Overstress Cumulative Time of Faced-Slab

(1) Dynamic stresses of faced-slab

Figure 6.11 shows the distributions of stress at 6 s moment, post-seismic stress and maximum stress (positive for tensile stresses and negative for compressive stresses) obtained based on a single sample for PGA = 0.4 g. Figure 6.12 shows the corresponding distribution of mean values. In terms of individual samples, the bottom of the faced-slab is mainly subjected to compressive stresses, and tensile stresses are mainly concentrated in the upper part of the faced-slab. However, from the mean value, the whole faced-slab is mainly subjected to compressive stress at the bottom, indicating that stresses are varying over time. Therefore, it is necessary to explore the damage standard based on the time accumulation effect.

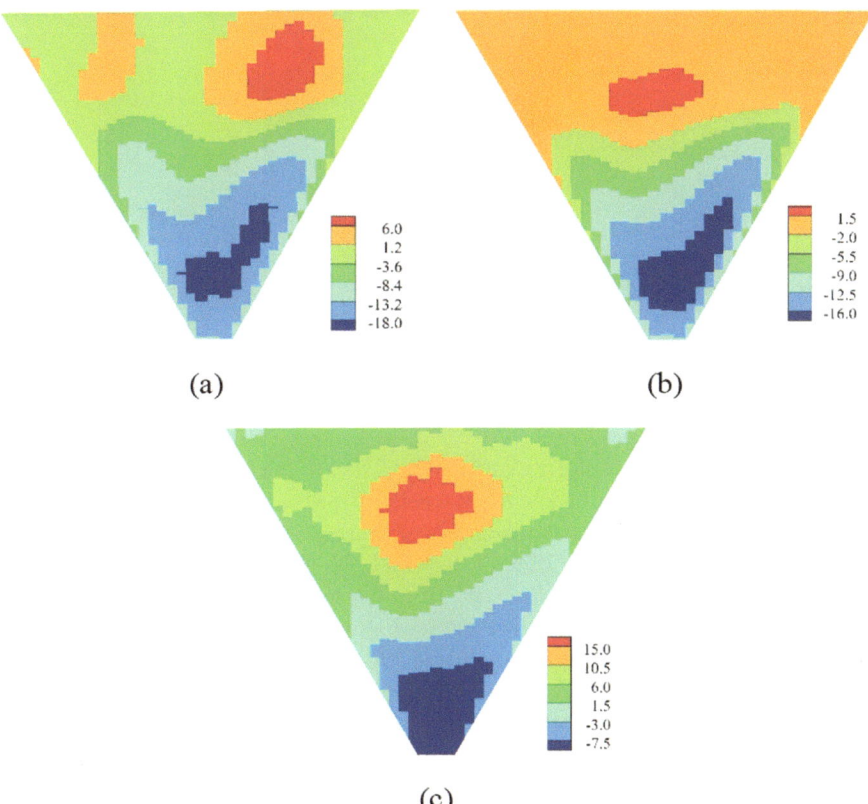

Fig. 6.11 Stress distribution in the slope direction of single sample at different time. **a** Stress at 6 s moment. **b** post-seismic stress. **c** maximum stress

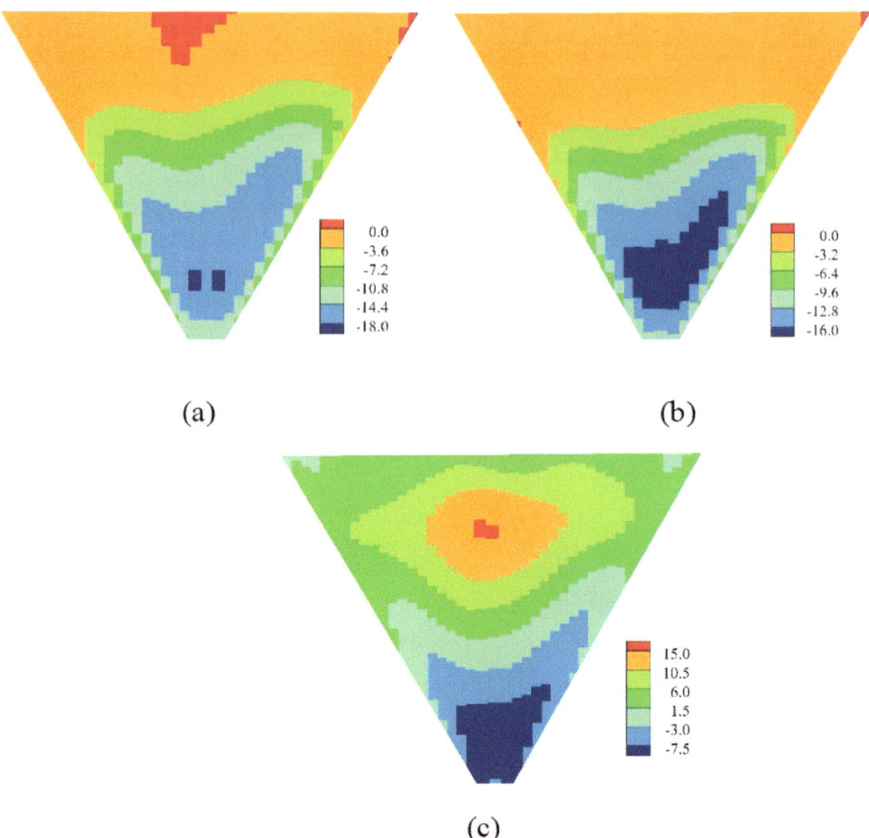

Fig. 6.12 Mean stress distribution in the slope direction at different time. **a** Stress at 6 s moment **b** Post-seismic stress. **c** Maximum stress

Figure 6.13 shows the stress distribution of slab in the slope direction along dam height under PGA = 0.4 g (positive for tensile stresses and negative for compressive stresses). The distribution pattern of stress response along dam height caused by different ground motions is different, in which the tensile stress is mainly concentrated at the top of the dam. However, from the mean value, the whole faced-slab is mainly subjected to compressive stress. Figure 6.13b shows the distribution of maximum stress along the dam height, which is basically similar to 2-D analysis. The distribution law of the response caused by different ground motions is basically same, but there is a big difference in the values, reflecting the necessity of analyzing based on the stochasticity of ground motions.

The four characteristic elements shown in Fig. 6.14 are selected to explore the stochastic dynamic response of the faced-slab during the seismic process in detail. Figure 6.15 shows the distribution pattern of the faced-slab in the slope direction along dam axial direction at different moments under PGA of 0.4 g (height of point

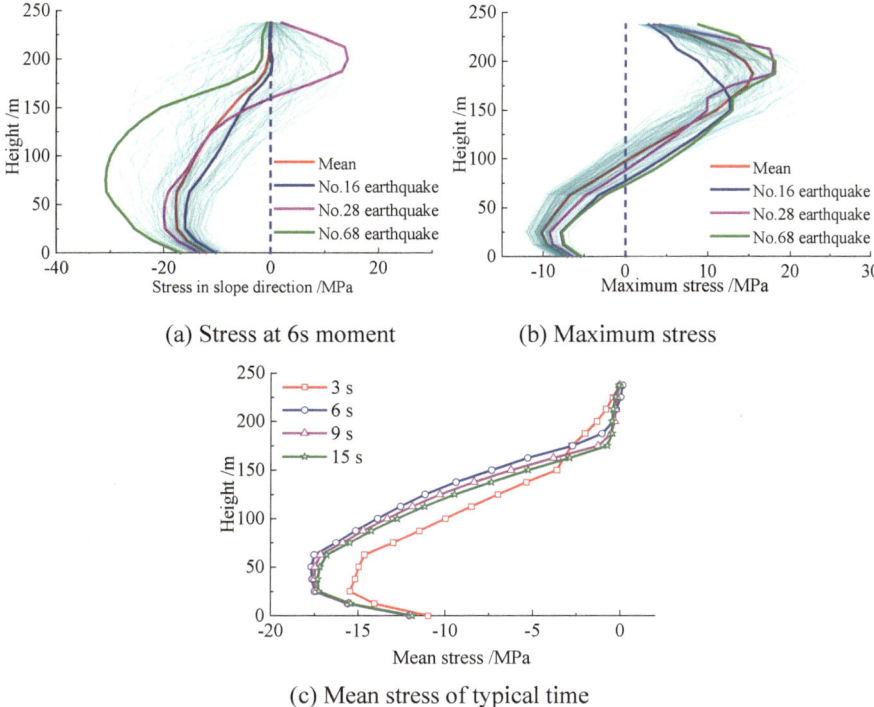

(a) Stress at 6s moment   (b) Maximum stress

(c) Mean stress of typical time

Fig. 6.13 Stress distribution in the slope direction along dam height. **a** Stress at 6 s moment **b** Maximum stress **c** Mean stress at typical time

C). The distribution pattern of downslope stresses under different ground motions varies greatly. Overall, the faced-slab stress tends to zero value at all moments, and the maximum stress is distributed at the center of the valley. Figure 6.16 shows the stress variation time course of the four feature elements, the tensile stresses are mainly distributed in the upper region of the faced-slab and the lower part of the faced-slab is mainly subjected to compressive stresses. From the mean value, the stress time course tends to be constant. From the standard deviation, the patterns at the top and bottom of the dam are closer, demonstrating a certain symmetry in the values. In summary, it is necessary to study the stress response law of faced-slab from the perspective of ground motion stochasticity.

(2) Overstress volume ratio

To represent the damage of the concrete slab more intuitively, this paper defines the ratio of the volume of tensile stress exceeding the tensile strength to the whole slab at different seismic strengths as the overstressed volume ratio (VolR). Figure 6.17 shows the VolR time history under different PGA, for example. Figure 6.18 shows the mean and the standard deviation time history. The VolR varies significantly with time evolution and seismic intensity, with an increasing trend for increasing

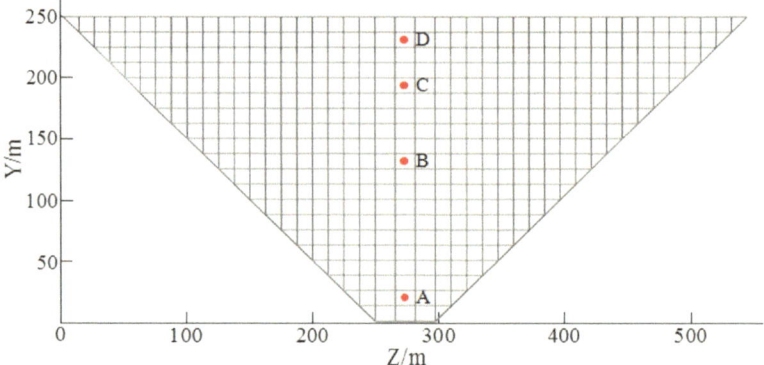

Fig. 6.14 Sketch of feature points locations

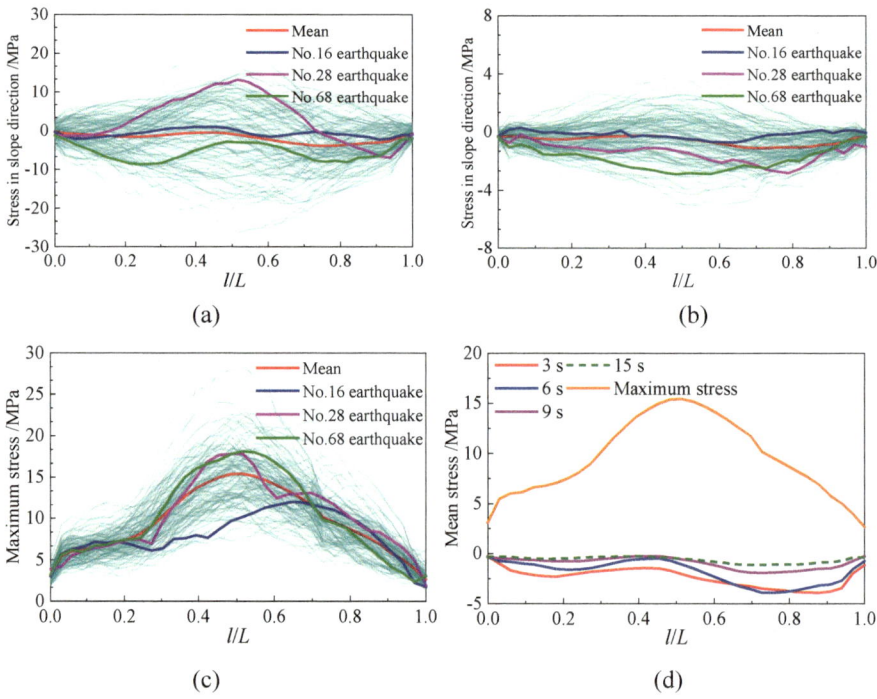

Fig. 6.15 Stress distribution in the slope direction along dam axial direction. **a** Stress at 6 s moment **b** Post-seismic stress **c** Maximum stress **d** Mean stress of typical time

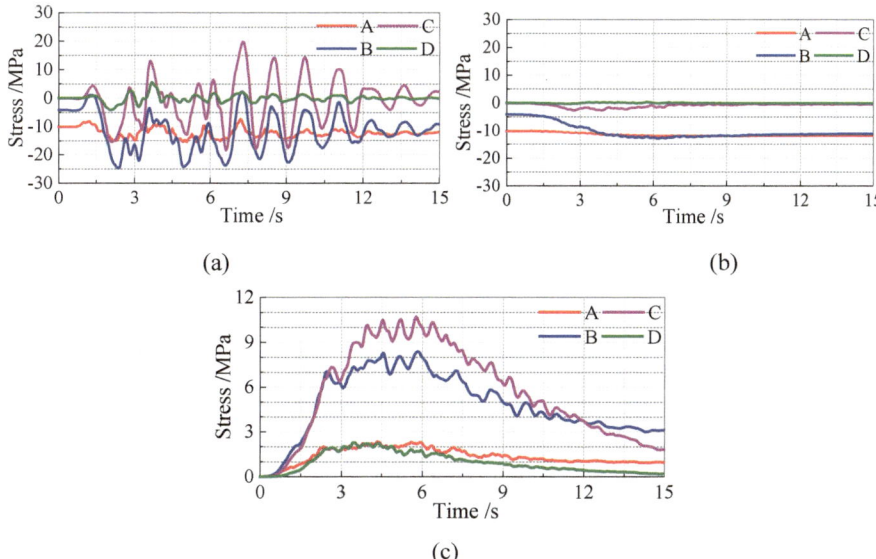

Fig. 6.16 Stress time history in the slope direction. **a** Typical sample **b** Mean **c** Standard deviation

seismic intensity. Therefore, considering the effect of time accumulation, the damage standard of faced-slab based on the VolR and overstress accumulation time can be initially explored.

Figure 6.19 illustrated the probability density evolution information of the overstressed volume ratio, including PDF curves of typical times, the PDF surfaces and contours of the overstressed volume ratio over the time from 8 to 12 s. The great variability and multi-peak shape shown in Fig. 6.19a greatly affects the dynamic reliability of faced-slab. Figure 6.19b, c show the irregular surface of PDF and the transfer process of probability information, proving the overstress volume ratio has great variability with the evolution of time.

Figure 6.20 shows the discrete point distribution and exceedance probability curves of different overstress volume ratios corresponding to cumulative time under PGA = 0.5 g, which can define the corresponding damage standard of faced-slab.

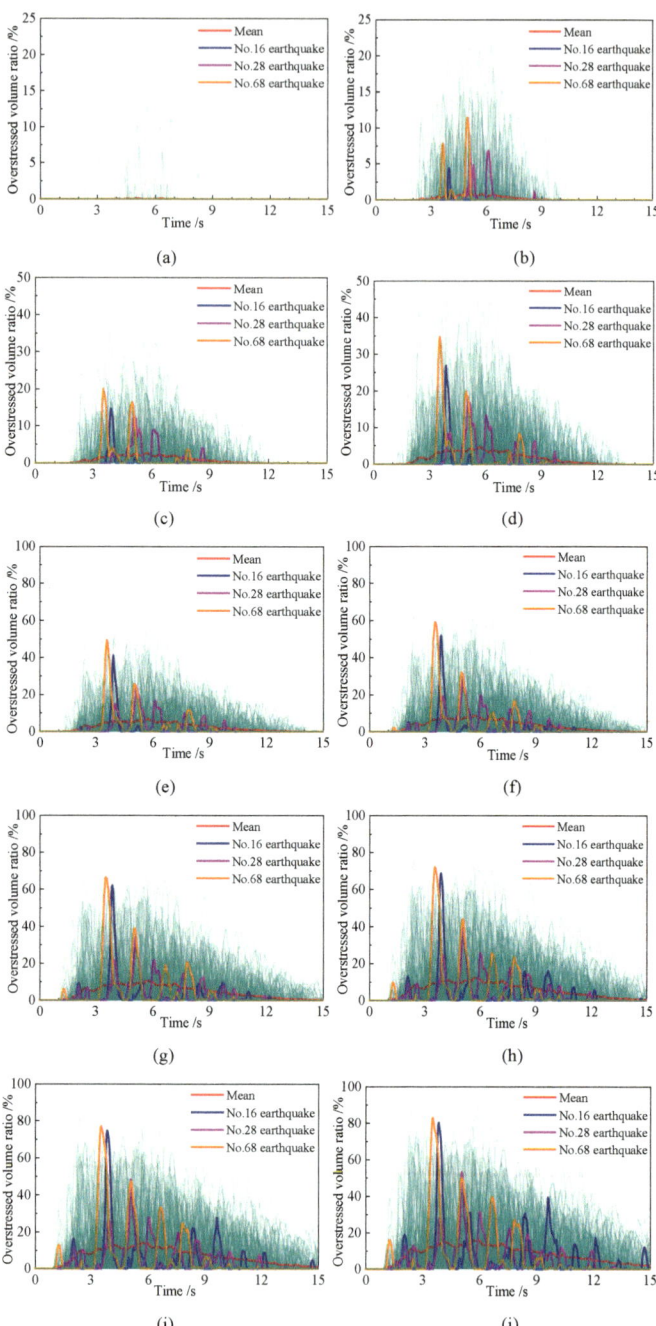

Fig. 6.17 Time history of overstress volume ratio under different PGA. **a** PGA = 0.1 g **b** PGA = 0.2 g **c** PGA = 0.3 g **d** PGA = 0.4 g **e** PGA = 0.5 g **f** PGA = 0.6 g **g** PGA = 0.7 g **h** PGA = 0.8 g **i** PGA = 0.9 g **j** PGA = 1.0 g

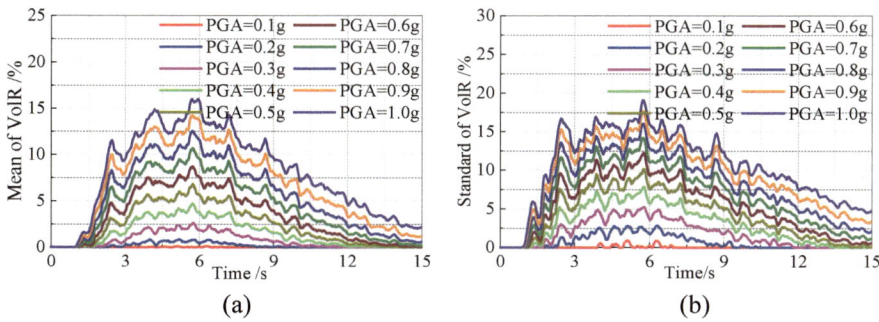

Fig. 6.18 Mean and standard deviation time history of overstress volume ratio under different PGA. **a** Mean **b** Standard deviation

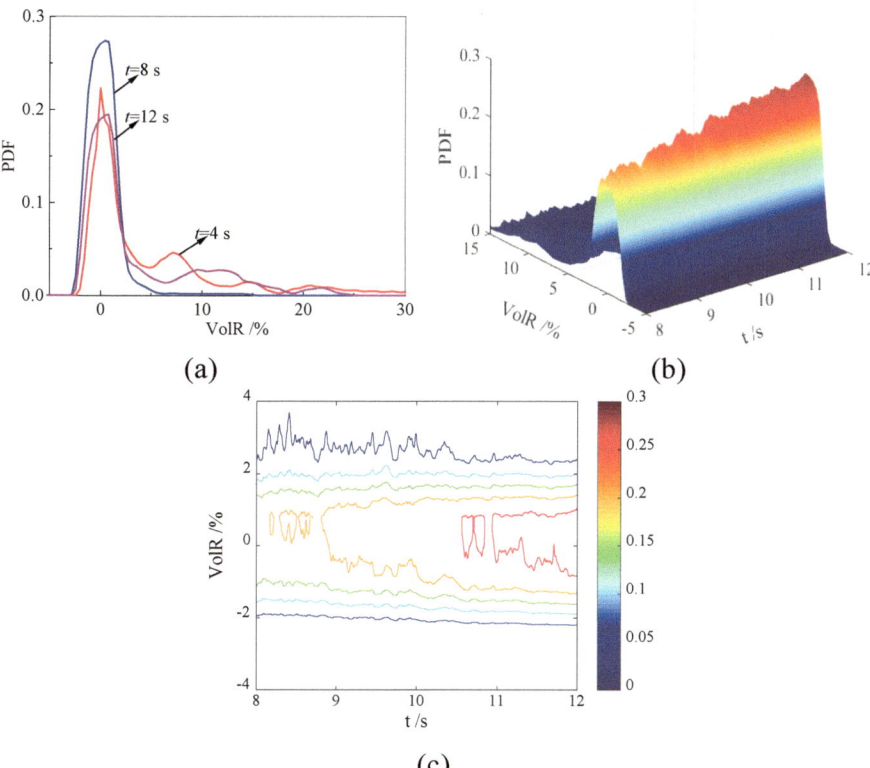

Fig. 6.19 Probability evolution information of overstress volume ratio. **a** PDF curves of typical times **b** PDF surfaces **c** Contour plot of PDF surface

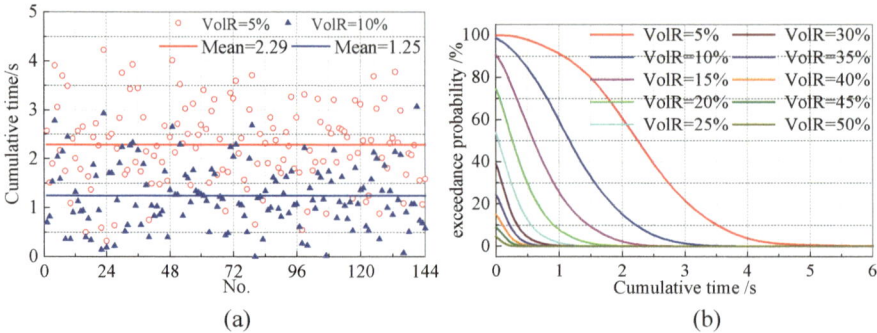

Fig. 6.20 Cumulative time and exceedance probability of different overstress volume ratios. **a** Discrete point distribution. **b** Exceedance probability curves

6.4 Conclusion

In this chapter, the stochastic dynamic response and probabilistic evolution of acceleration, deformation and faced-slab stress of a three-dimensional high CFRD are analyzed with full consideration of the randomness of ground motions, and the range of distributional variations of the physical quantities is suggested from the stochastic dynamic and probabilistic perspectives by combining the stochastic ground motion generation method, the generalized probability density evolution method, the fragility analysis method, and the elasto-plasticity analysis method of CFRDs. Finally, the performance safety evaluation standards based on the relative seismic subsidence rate of the dam crest are compared, and the faced-slab seismic safety evaluation indicators based on the overstress volume ratio combined with the overstress accumulation time is preliminarily explored. The main work and conclusions are as follows:

(1) From the perspective of stochastic dynamics, it is revealed that the acceleration and deformation of the dam body are mainly distributed in the central part of the valley at the top of the dam, and the top area of the dam shows strong "whip sheath" effect and three-dimensional effect. From the perspective of stochastic dynamics and probability, the acceleration, deformation and stress response have large variability, both in distribution and value. It is necessary to analyze the non-linear dynamic response of high CFRDs under earthquake action from the stochastic dynamics point of view. The result has certain reference significance for the seismic safety evaluation and ultimate seismic capacity analysis of high CFRDs.

(2) From the perspective of probability and performance evaluation, the damage probability relationship between 2-D deformation and 3-D deformation is revealed; the seismic safety evaluation framework based on the overstressed volume ratio of faced-slabs combined with the cumulative time is preliminarily explored and suggested, as described in Sect. 8.1, which basically indicates the reasonableness of the division of the performance level through the probability, and basically corresponds to the deformation-based performance safety evaluation standard, and further improves the performance-based seismic safety evaluation framework for CFRDs.

It should be noted that for panels, it is most reasonable to establish the corresponding seismic performance safety evaluation standards based on elastic–plastic cumulative damage, but the current numerical calculations based on the 3D concrete elastic–plastic damage model are particularly large in scale and are not well suited for stochastic dynamic and probabilistic analyses. Nevertheless, the performance indicators established in this chapter can basically quantify the damage degree of faced-slabs, which provides a reference for performance-based seismic safety evaluation of CFRDs.

Reference

Goodman RE (1968) A model for the mechanics of jointed rock. J Soil Mech Found Div. In: Proceedings of ASCE

Chapter 7
Stochastic Seismic Analysis and Performance Safety Evaluation for Slope Stability of High CFRDs

7.1 Introduction

Numerous engineering cases (Guan 2009; Liu et al. 2015), dynamic numerical analysis (Zou et al. 2013; Uddin 1999) and dynamic physical model tests (Zhu et al. 2011; Liu et al. 2016) demonstrated that the instability of dam slopes is a major engineering concern for high CFRDs under earthquake excitation. It is explicitly mandated that seismic stability calculations for earth-rock dams should be included in seismic analysis in China's Hydraulic Seismic Design Code (NB 35047–2015). The comprehensive evaluation of slope stability is further specified to consider the factors such as the position, depth, and extent of the slip surface, as well as the duration and magnitude of stability index exceeding limits. Therefore, in addition to the traditional safety factor, the cumulative time of $F_S < 1.0$ and cumulative slippage were adopted by many scholars to evaluate the stability of dam slopes (Zhang and Li 2014). This approach, along with associated performance indices, was commonly utilized in numerous engineering practices. The seismic response of dam slope is characterized by a multitude of uncertainties, primarily including the stochastic seismic excitation and the material parameter randomness. However, deterministic dynamic time history analysis based on one or several seismic waves is more commonly employed in the current methods of dam slopes seismic performance assessment. The current approach fails to quantitatively analyze the impact of uncertainty factors on the seismic stability of dam slopes and inadequately evaluates the seismic safety of dam slopes from a performance perspective. Therefore, studying dam slope seismic stability based on stochastic dynamic and probabilistic analysis can effectively supplement deterministic analysis methods by fully considering uncertain factors in seismic response processes and introducing reasonable performance indices for dam slope stability evaluation. This approach is of significant importance and can progressively enhance performance-based seismic stability assessments of high CFRD slopes.

© The Author(s) 2025
B. Xu and R. Pang, *Stochastic Dynamic Response Analysis and Performance-Based Seismic Safety Evaluation for High Concrete Faced Rockfill Dams*, Hydroscience and Engineering, https://doi.org/10.1007/978-981-97-7198-1_7

In addition, some studies have shown that instability of dam slopes tends to be shallow layer sliding at the top of the dam body, with slip surfaces often occurring under low confining pressures. The softening effects of rockfill materials are gradually manifested under low confining pressures subjected to earthquakes (Zou et al. 2016), especially strong ones and this will exacerbate damage to the dam slopes (Skempton 1985). Hence, it is of great significance to analyze the seismic performance of the high CFRD slopes considering the softening effects and some researchers have conducted a lot of studies. For instance, Chen et al. (1992) conducted finite element analysis by using strain softening model to determine the degree of weakening along potential slip surfaces. Potts et al. (1990) conducted numerical simulations using a combination of finite element method and strain softening model to investigate the delayed failure and progressive cumulative failure of an excavated stiff clay slope. Zhang et al. (2007) proposed a methodology for analyzing the progressive failure of strain softening slopes based on the strength reduction method and strain softening model. Liu and Ling (2012) investigated the effects of strain softening of backfill on the deformation and reinforcement load of geosynthetic-reinforced soil structures (GRS) walls. Wang et al. (2017) considered the strain-softening characteristics of soils and simulated the mechanical behavior of slope failure using an improved finite element method. These studies indicate that if the strain softening characteristics of the slope are neglected, the slope safety might be overestimated, particularly under earthquakes. However, few studies have been conducted to consider the impact of rockfill softening characteristics on the seismic safety of dam slopes. Zhou et al. (2016) performed a deterministic comparison of the influence on the safety of dam slopes by considering and without considering softening characteristics of rockfill materials, and preliminary findings suggested that the potential for dam slope instability increased when considering the softening effects. However, deterministic analyses fall short in providing a comprehensive reflection of the stochastic ground motions and the effects of different seismic intensities on the stability of dam slopes. Hence, conducting thorough research on the impact of rockfill material softening effects on the seismic safety of dam slopes from a stochastic and probabilistic perspective is imperative. Additionally, the conventional dam slope stability evaluation methods relying on the pseudo-static approach have limitations in accurately portraying the input characteristics of ground motions and their dynamic responses, which significantly influence dam slope stability. Therefore, employing finite element dynamic time-history analysis methods is necessary for a comprehensive assessment of dam slope seismic stability. Furthermore, selecting appropriate performance indices to establish corresponding performance-based seismic safety evaluation criteria is crucial.

In this chapter, the randomness of material parameters, stochastic seismic excitation, and their coupling are comprehensively taken into account. The dynamic finite element time-history method for dam slope stability analysis was applied, incorporating the coupled calculation method for softening strength parameters variation of the rockfill materials. A methodology for generating stochastic ground motions and random parameters was established and the fragility analysis as well as the GPDEM were introduced. The impact of rockfill material softening effects on the seismic stability of dam slopes is evaluated from a stochastic and probabilistic perspective based on three performance indices: safety factor, cumulative time of $F_S < 1.0$ and cumulative slippage, and the stochastic dynamic response pattern of dam slopes stability is revealed. Furthermore, a probabilistic analysis method for dam slope stability considering the softening effects is established, along with a performance-based seismic safety evaluation framework of high CFRD slopes.

7.2 Dynamic Finite Element Time History Analysis Method Considering Softening Behaviors of Rockfill for Dam Slope Stability

7.2.1 Dynamic Finite Element Time History Analysis Method Considering Softening Behaviors of Rockfill for Dam Slope Stability

Although the pseudo-static method is simple and has rich practical experience, it fails to adequately consider the stress–strain relationship within the soil mass. The calculation results only represent the average values of assumed potential sliding surfaces and cannot provide information on the magnitude of soil deformation. Therefore, methods based on seismic response analysis have gradually gained attention and development. A method that calculates the factor of safety against sliding for each moment by considering the instantaneous stress changes during seismic events is referred to as the finite element time-history method.

The safety factor can be expressed as follows:

$$F_s = \frac{\sum_{i=1}^{n} (c_i + \sigma_i \tan\varphi_i) l_i}{\sum_{i=1}^{n} \tau_i l_i} \tag{7.1}$$

in which, c_i and φ_i are the cohesion and internal friction angle of element i, respectively, and l_i is the length of the element i in the slip circle. σ_i and τ_i are the normal stress and tangential stress of element i, respectively, and as expressed follows:

$$\sigma = \frac{\sigma_x + \sigma_y}{2} - \frac{\sigma_x - \sigma_y}{2}\cos2\alpha - \tau_{xy}\sin2\alpha \tag{7.2}$$

$$\tau = \frac{\sigma_x - \sigma_y}{2}\sin2\alpha - \tau_{xy}\cos2\alpha \tag{7.3}$$

where, $\sigma_x = (\sigma_x^s + \sigma_x^d)$, $\sigma_y = (\sigma_y^s + \sigma_y^d)$; $\tau_{xy} = (\tau_{xy}^s + \tau_{xy}^d)$; σ_x^s is the horizontal static stress; σ_x^d is the horizontal dynamic stress; σ_y^s is the vertical static stress; σ_y^d is the vertical dynamic stress; τ_{xy}^s is the static shear stress; and τ_{xy}^d is the dynamic shear stress; α denotes the inclination of the base of the slice with respect to the horizontal.

For any slip surface, the sliding angular acceleration of the slider can be expressed as:

$$\ddot{\theta}(t) = \frac{M}{I} \tag{7.4}$$

$$M = \left[\sum_{i=1}^{n} \tau_i l_i - \sum_{i=1}^{n} (c_i + \sigma_i \tan\varphi_i) l_i \right] R \tag{7.5}$$

where I is the moment of inertia, M is the torque and R is the sliding surface radius.

When an instantaneous slide occurs at a certain instant, the slippage of the slip circle will be:

$$D_i^k = R^k \theta_i^k = R^k \iint \ddot{\theta}_i^k \mathrm{d}t \tag{7.6}$$

Many instantaneous slides may occur throughout the time period, and the cumulative slippage will be:

$$D^k = \sum_{i=1}^{n} D_i^k \tag{7.7}$$

The maximum cumulative slippage of the dam slope can be expressed as:

$$D_{\max} = \max\left(D^1, D^2, \cdots, D^k, \cdots, D^m\right) \tag{7.8}$$

The schematic diagram of the safety factor calculation is show in Fig. 7.1.

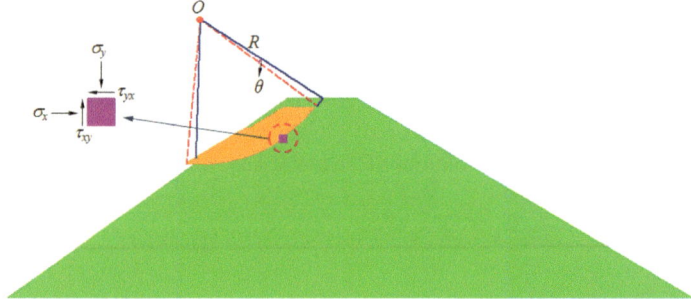

Fig. 7.1 Schematic diagram of the safety factor calculation

7.2.2 Rockfill Softening

In the currently commonly used block slip method, the decrease in shear strength of the slip surface is not considered when $F_S<1.0$. However, the peak strength of the rockfill materials gradually decreases along with the sliding. In this section, the software FEMSTABLE 2.0 developed by Dalian University of Technology was adopted to perform stochastic dynamic response analysis of the dam slope and the softening characteristic of the rockfill materials was considered. The effect of stochastic ground motions on the minimum safety factor, cumulative time of $F_S<1.0$ and cumulative slippage was thoroughly investigated. The specific steps are described as below:

1. Five sets of ground motions were generated using the stochastic ground motion model. The PGA was adjusted from 0.1 g to 0.5 g in intervals of 0. 1 g and each set had 233 stochastic ground motions.
2. For each ground motion, the static and dynamic calculations of the dam are conducted, and the finite element time-history analysis of the dam slope is performed based on the accumulated static and dynamic stress results, then the safety factor of the dam slope is obtained.
3. The cumulative slippage of the slip surface was calculated using the block slip method if the safety factor was less than 1.0. The shear strain on the slip surface can be expressed as:

$$\gamma_s = \frac{D^k}{d} \tag{7.9}$$

where D^k is the cumulative slippage of the slip block, d is the shear band width and γ_S is the average shear strain.

The relationship between shear strain and post-peak strength is determined using the stress–strain curve of the rockfill materials. If the minimum safety factor is less than 1.0, the subsequent finite element dynamic stability analysis is conducted using the post-peak strength, conversely, using the peak strength.

4. Check if the earthquake had ended. If the earthquake was ongoing, continue the calculation. If the earthquake had ended, output the time history of safety factor and cumulative slippage. Repeat this process for each ground motion and a series of dynamic stability information under different earthquake intensities were obtained.
5. The GPDEM were introduced and the probabilistic information of safety factor, cumulative time of $F_S< 1.0$ and cumulative slippage under different seismic intensities were obtained. The effects of considering or without considering the softening effects on the stability of the dam slopes were analyzed from a probabilistic perspective.

7.3   Effect of Softening Characteristics of Rockfill Materials Based on Stochastic Dynamic and Probabilistic Analysis

A 250-m CFRD was used to perform finite element time history analysis of dam slope stability in order to compare the effects on the slope stability of high CFRDs considering and without considering softening. First, five sets of stochastic ground motions were generated from PGA = 0. 1 g to 0. 5 g with 0. 1 g intervals and each set had 233 samples. Then, for each ground motion, the 2-D nonlinear finite element numerical analysis of the CFRD was performed including static, dynamic and stability calculations using the software GEODYNA and FEMSTABLE 2.0. Finally, the influence of softening effects of rockfill materials was evaluated based on the minimum safety factor, cumulative time of $F_S < 1.0$ and cumulative slippage from the perspective of stochastic dynamics and probability, so as to the basis and reference of performance-based seismic safety evaluation for high CFRD slopes.

7.3.1   Calculating Basic Information

The calculation models, load conditions and boundary conditions remain consistent with those described in Sect. 2.7.4, and the static and dynamic parameters are also adopted from the numerical values presented in Sect. 2.7.4. An improved Newmark method was adopted to perform the stability calculations, considering and without considering softening. Ten groups of consolidated drained triaxial tests results of rockfill materials from existing or planned high earth-rock dams were statistically analyzed, and the relationship between post-peak shear strain of rockfill materials and dimensionless values (ratio of normalized post-peak strength to peak strength) was obtained, also a normalized strength parameter curve was fitted. Figures 7.2 and 7.3 illustrate the relationships between $\frac{\varphi_0}{\varphi_{max}}$ and $\frac{\Delta\varphi}{\Delta\varphi_{max}}$ with post-peak shear strain, respectively. As can be seen, when the post-peak shear strain increases, the softening effects of the rockfill materials become evident. The relationship equation obtained from the fitting curves are Eqs. (7.10) and (7.11). Therefore, the relationship between post-peak shear strain and post-peak strength parameters used for the CFRD can be obtained, as shown in Table 7.1. The softening characteristic was considered only in rockfill A and B in this paper.

$$\varphi_0/\varphi_{max} = -0.6315 \cdot \gamma_s + 1 \tag{7.10}$$

$$\Delta\varphi/\Delta\varphi_{max} = -2.2617 \cdot \gamma_s + 0.9938 \tag{7.11}$$

Fig. 7.2 Normalized relationships between γs and φ₀/φmax

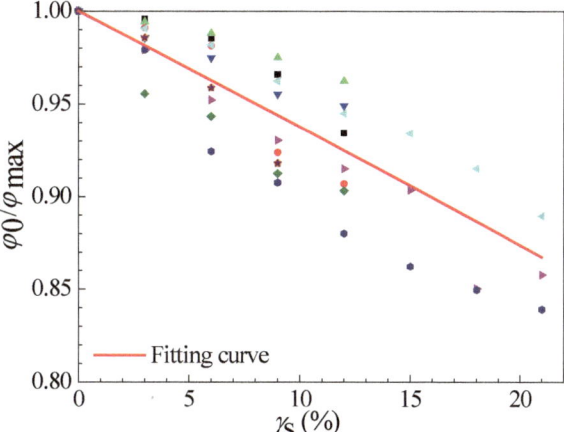

Fig. 7.3 Normalized relationships between γs and Δφ/Δφmax

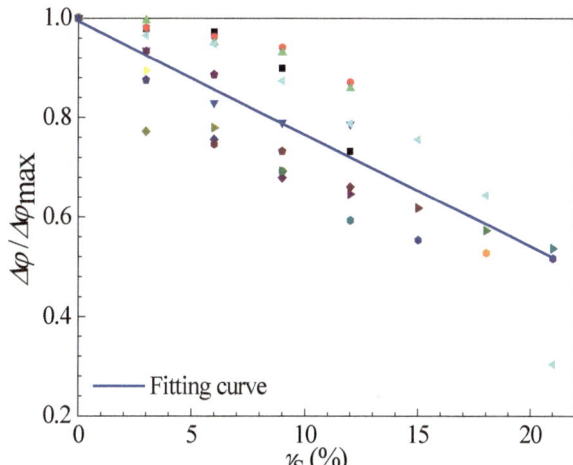

Table 7.1 Strength after peak value

Post-peak shear strain	Rockfill A and B	
	$\varphi_0(°)$	$\Delta\varphi(°)$
0	49.80	7.20
0.03	48.85	6.65
0.06	47.96	6.16
0.09	47.01	5.67
0.12	46.07	5.18
0.15	45.12	4.69
0.18	44.17	4.20
0.21	43.20	3.72

7.3.2 Analysis of Calculating Results

The stochastic dynamic results of three indices (the minimum safety factor, cumulative time of $F_S < 1.0$ and cumulative slippage) for dam slope stability under different PGA (0. 1 g, 0.2 g, 0.3 g, 0.4 g, 0. 5 g) considering and without considering softening were obtained. The probabilistic information of these three indices was also obtained based on the GPDEM and equivalent extreme event theory.

(1) Safety factor

Figure 7.4 illustrates the second-order statistical time histories (mean and standard deviation) of the safety factor with PGA adjusting from 0. 1 g to 0. 5 g. The individual safety factor time history analysis suggests that the impact of considering softening on the safety factor is not obvious. However, the mean and standard deviation of the safety factor indicate that considering softening effects leads a decrease of the safety factor. There is little difference in the safety factor time history between considering and without considering the softening effects under weak earthquake (e.g., PGA = 0.1 g), which is because the rockfill material has not reached the peak strength to show the softening effects. However, as the seismic intensity increases (e.g., PGA = 0.5 g), the difference becomes more obvious. Furthermore, the difference in the safety factor between the two cases gradually increases with the increase of seismic intensity, indicating a progressive trend. The standard deviation increases with time, indicating that the change of safety factor increases with the development of nonlinear behavior of rockfill materials which is further illustrated by the sharp fluctuation of 3–9 s; the standard deviation also decreases over time, part of the reason is that the change of safety factor decreases as the history of ground motion time decreases with time. This indicates that the softening of the rockfill materials is a progressive process. Furthermore, the stability of the mean and standard deviation of the safety factor time history demonstrates the statistical significance of the generated stochastic ground motions. It also indicates that the randomness of the ground motions has a significant impact on the safety factor. Based on the above analysis, it is evident that the dynamic response of dam slope stability is sensitive to the stochastic ground motions and different ground motions have a significant impact on the safety factor of the dam slope. The softening of the rockfill materials under earthquake is a progressive process. Therefore, it is necessary and meaningful to consider the softening of the rockfill materials and introduce the stochastic ground motions for the analysis of dam slope stability based on safety factor.

Figures 7.5 and 7.6 illustrate the PDFs and CDFs of the minimum safety factor with PGA = 0. 2 g and 0. 5 g based on the equivalent extreme event theory. It is evident that there is negligible difference in the PDFs and CDFs of the minimum safety factor with PGA = 0.2 g between considering and without considering softening. However, the difference increases significantly with PGA = 0. 5 g and it is consistent with the results of mean and standard deviation. The results show that as the PGA increases, the safety factor time history and the minimum safety factor of the dam slope show an increasing difference between considering and without considering softening during

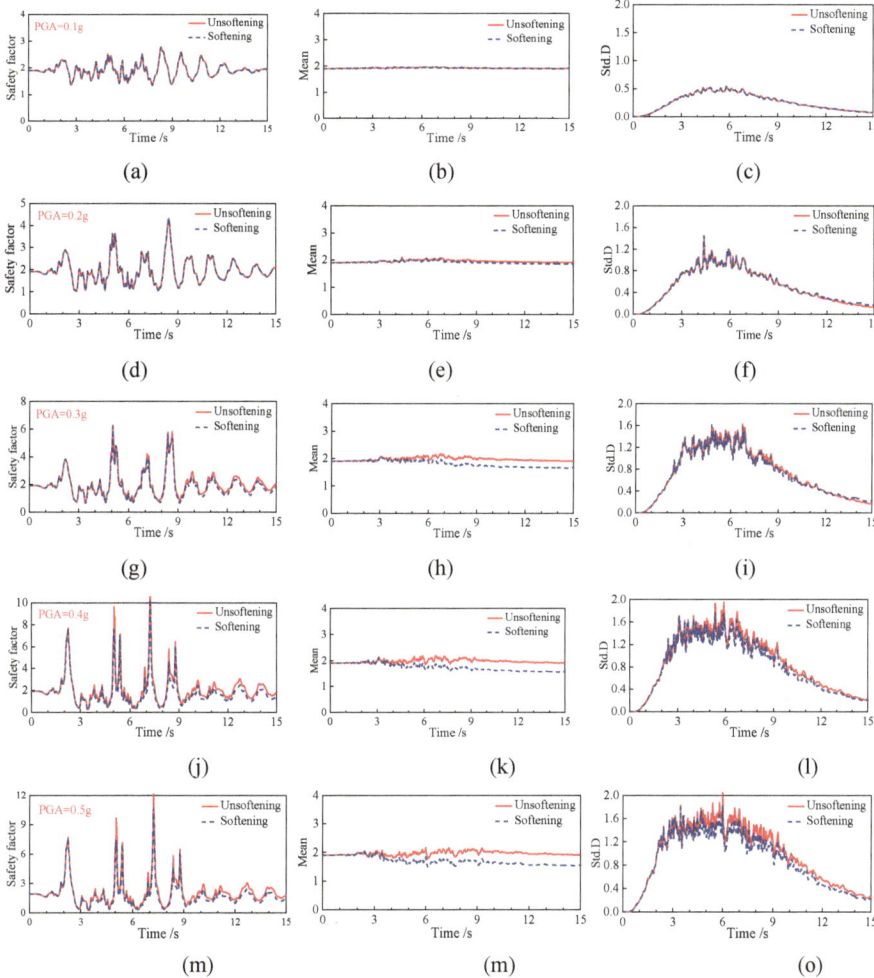

Fig. 7.4 Safety factor time history and their mean and standard deviation. **a** Safety factor. **b** Mean **c**. Standard deviation **d**. Safety factor **e**. Mean **f** Standard deviation **g**. Safety factor **h**. Mean **i**. Standard deviation **j**. Safety factor **k**. Mean **l**. Standard deviation **m**. Safety factor **n**. Mean **o**. Standard deviation

the earthquake process. This difference is attributed to the progressive softening effects of the rockfill materials with the increasing seismic intensity. The dam slope reliabilities with the minimum $F_S = 1.0$ are listed in Table 7.2. It can be observed that there is almost no difference in probability between considering and without considering softening corresponding to different earthquake intensities. This result also indicates that it is unreasonable to evaluate the dam slope stability only based on the minimum safety factor.

(2) Cumulative time of $F_S < 1.0$

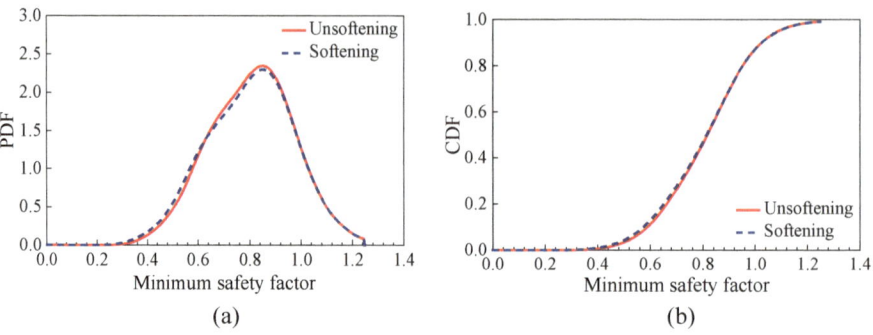

Fig. 7.5 PDFs and CDFs of minimum safety factor under 0.2 g (**a**) PDF (**b**) CDF

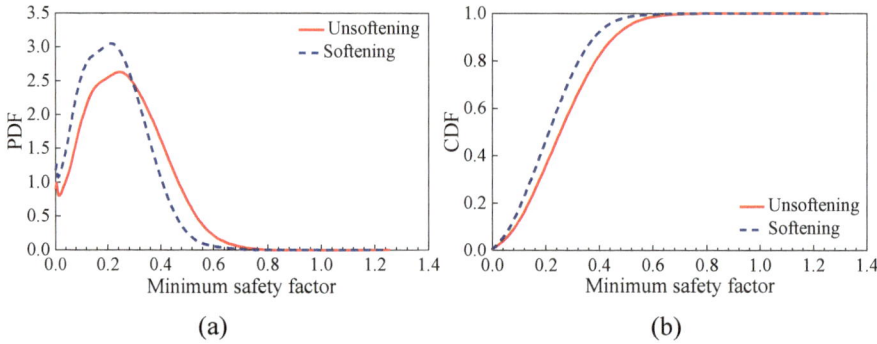

Fig. 7.6 PDFs and CDFs of minimum safety factor under 0.5 g. **a**. PDF **b** CDF

Table 7.2 Seismic reliability under different earthquake levels

Softening effects	Cumulative time (s)	PGA				
		0.1 g	0.2 g	0.3 g	0.4 g	0.5 g
Unsoftening	0	0.9738	0.0931	0.0043	0	0
	1	1	0.9828	0.6314	0.1226	0.0199
	2 `	1	1	0.9977	0.9132	0.5685
Softening	0	0.9738	0.0931	0.0043	0	0
	1	1	0.9355	0.3314	0.0274	0
	2	1	1	0.8661	0.3132	0.0593

Note 0 represents the strain before peak value

The cumulative time of $F_S < 1.0$ is a new indicator for the dam slope stability evaluation, which has been gradually adopted by researchers. Some scholars and engineers suggest that the dam slope will lose stability if the cumulative time of $F_S < 1.0$ exceeds 1–2 s. Figures 7.7 and 7.8 illustrate the PDFs and CDFs of the cumulative time of $F_S < 1.0$ with PGA $= 0.2$ g and 0.5 g based on the equivalent extreme event theory, respectively. It can be observed that under weak earthquake, there is little difference between considering and without considering softening. However, as the seismic intensity increases, the difference becomes more obvious, indicating that the softening effects of the rockfill materials gradually become evident under strong earthquake actions. Furthermore, the difference in CDFs between considering and without considering softening gradually increases after the cumulative time of 0.5 s with PGA $= 0.2$ g. This is because once the rockfill materials became softening, the post-peak strength decreases, leading to lower calculated safety factors and an increase in the cumulative time of $F_S < 1.0$. The reliability based on the cumulative time of $F_S < 1.0$ corresponding to different earthquake intensities are listed in Table 7.2. The results show that as the seismic intensity increases, the difference in probabilities between considering and without considering softening becomes more obvious and this also suggests that evaluating the stability of the dam slope based on the cumulative time of $F_S < 1.0$ is reasonable.

(3) Slip surfaces and cumulative slippage.

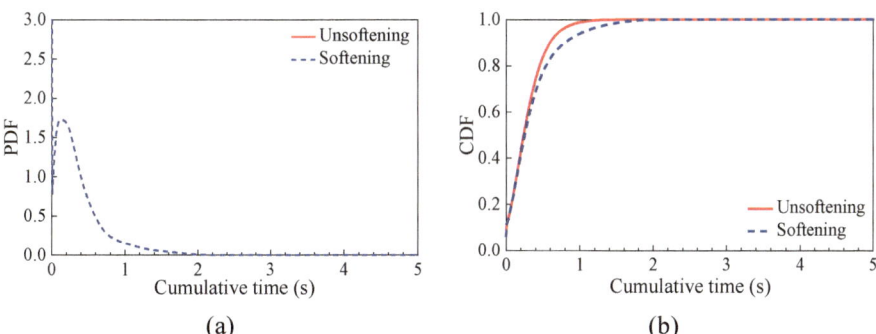

Fig. 7.7 Probability information of cumulative time of $F_S < 1.0$ under 0.2 g (**a**) PDF (**b**) CDF

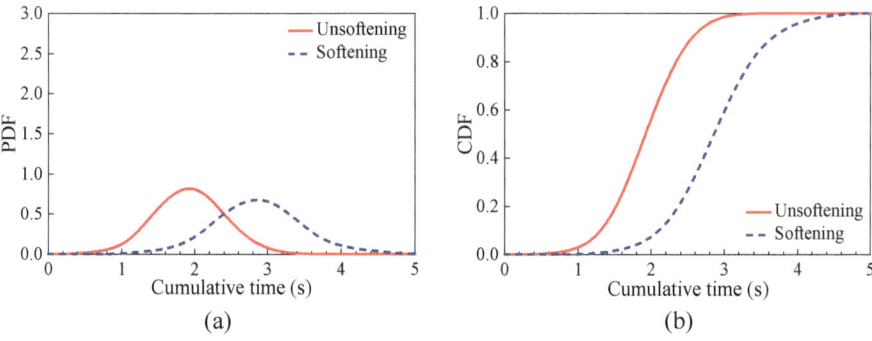

Fig. 7.8 Probability information of cumulative time of $F_S < 1.0$ under 0.5 g

Figure 7.9 illustrates the most dangerous slip surfaces of dam slope with PGA = 0. 5 g considering and without considering softening. The results show that the positions of the most dangerous slip surfaces are basically not affected by considering and without considering softening corresponding to different earthquakes, but the positions of some slip surfaces are also different, which proves that different earthquake motions have certain influence on the position of the most dangerous slip surface, and shows the necessity of studying the influence of softening effects on dam slope slip from the perspective of stochastic dynamics and probability. Figure 7.10 illustrates the cumulative slippage with PGA = 0. 5 g considering and without considering softening. It is evident that the cumulative slippage considering softening is significantly greater than that without considering softening under strong earthquake. Figures 7.11 and 7.12 illustrate the probability information with PGA = 0. 2 g and 0. 5 g considering and without considering softening. Table 7.3 list the reliability of the cumulative slippage of 5, 50, and 100 cm corresponding to different PGA, both considering and without considering softening. The probability information demonstrates that under weak earthquake, the dam slope exhibits almost no slippage, while significant slippage occurs and the softening effect has a substantial impact on the slippage under strong earthquake. This is because with the earthquake intensity increases, the softening effects becomes more obvious and the slippage and slip shear strain also increase.

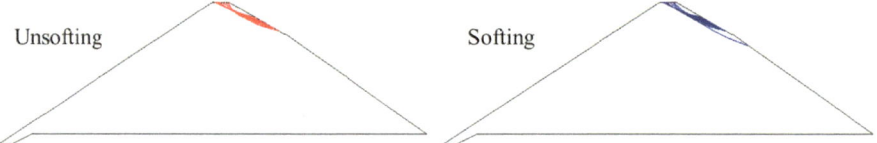

Fig. 7.9 The slide surface corresponding to the minimum safety factor (0. 5 g)

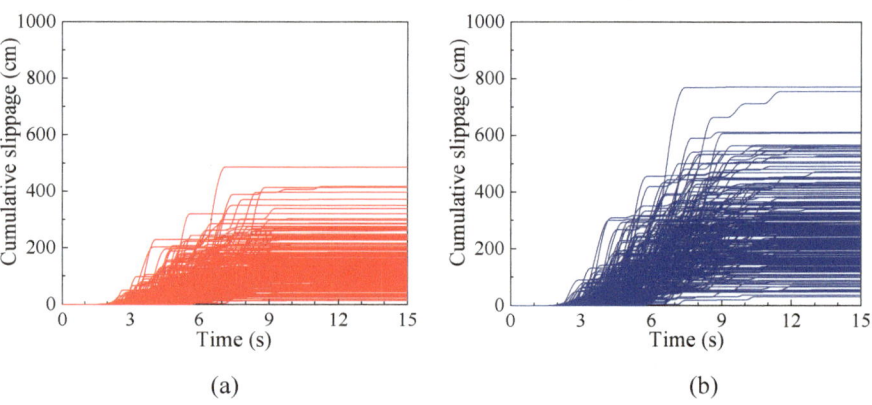

Fig. 7.10 The cumulative slippage of dam slope corresponding to the minimum safety factor (0. 5 g) (**a**) Unsoftening (**b**) Softening

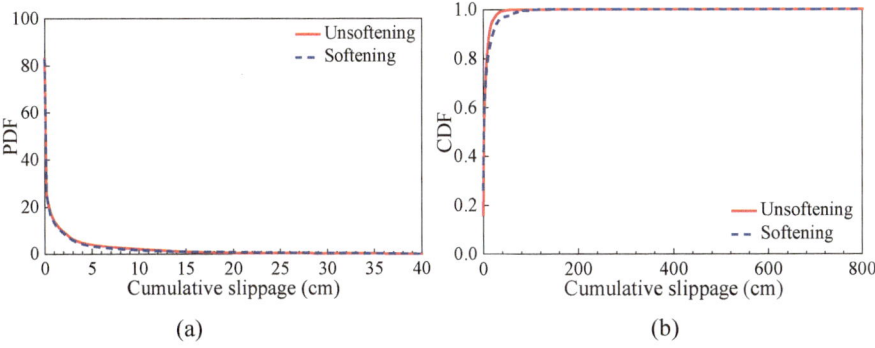

Fig. 7.11 Probability information of cumulative slippage under 0.2 g (**a**) PDF (**b**) CDF

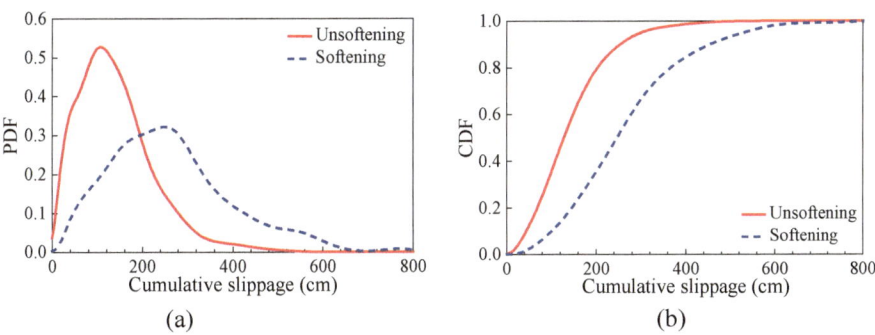

Fig. 7.12 Probability information of cumulative slippage under 0.5 g. **a** PDF. **b** CDF

Table 7.3 Seismic reliability under different earthquake levels

Softening effects	Cumulative slippage (cm)	PGA				
		0.1 g	0.2 g	0.3 g	0.4 g	0.5 g
Unsoftening	5	1	0.7121	0.1340	0.0218	0
	50	1	0.9976	0.8299	0.3773	0.1266
	100	1	1	0.9734	0.7297	0.3522
Softening	5	1	0.6869	0.1067	0.0102	0
	50	1	0.9732	0.5378	0.1252	0.0240
	100	1	0.9969	0.8170	0.3305	0.0979

7.3.3   *Conclusion*

The rockfills of earth-rockfill dams gradually show softening effects subjected to earthquakes, especially strong ones, which will significantly affect the safety of dam slopes. In order to evaluate the softening effects on the stability of dam slopes, a probability analysis method of dam slope safety based on the equivalent extreme event theory and GPDEM was introduced considering the stochastic earthquake excitation. A 250-m CFRD was used to perform stochastic dynamic response analysis and probabilistic reliability analysis based on three physical parameters of dam slope stability, the minimum safety factor, cumulative time of $F_S < 1.0$ and cumulative slippage. Three main conclusions are as follows:

1. In this section, the fundamental idea of GPDEM was employed to construct a stochastic process with a "virtual time parameter" regarding the extreme dynamic responses of the dam slope. The GPDEM equation was derived and the extreme value distribution probability and reliability of stochastic structural dynamic reaction are obtained. This method exhibits wide applicability in the probabilistic analysis of complex engineering structures, enabling a more accurate assessment of the reliability of earth-rockfill dam slopes.
2. The results show that the difference between considering and without considering softening gradually increases with the increase of earthquake intensity based on the minimum safety factor, cumulative time of $F_S < 1.0$ and cumulative slippage which was because the softening effects of rockfills were gradually revealed during the earthquake. Meanwhile, the softening was a gradual process and it is of great significance to analyze the seismic performance of the high CFRDs considering the softening effects.
3. The results of reliability analysis show that it is unreasonable to study the stability of earth-rockfill dams only based on the minimum safety factor, and it is necessary to combine the cumulative time of $F_S < 1.0$ and cumulative slippage to fully evaluate the safety of dam slope. The proposed stochastic probabilistic analysis method can give a more accurate evaluation of the reliability of high earth-rockfill dam slopes and also provide theoretical support for seismic design and risk assessment of earth and rock dams.

7.4 Statistical Analysis of Shear Strength Parameters of High CFRDs

Because the rockfill materials of earth-rockfill dams are sourced from different quarries or are located in different areas, there exists a certain level of discreteness and variability. In order establish a universal, unified and referential safety evaluation system for seismic performance of high CFRD slopes, the shear strength parameters of 40 CFRDs with a height of more than 100 m are statistically collected to obtain exact parameter statistical characteristics, as shown in Table 7.4.

Figures 7.13 and 7.14 illustrate the frequency histograms and statistical features of φ_0 and $\Delta\varphi$, respectively. It is evident that φ_0 and $\Delta\varphi$ basically obey a normal distribution. The mean and standard deviation are listed in Table 7.4. The relationship between post-peak strength and post-peak strain based on the mean shear strength can be obtained according to Eqs. (7.10) and (7.11). The corresponding values are presented in Table 7.5.

Table 7.4   The shear strength parameter statistics of high CFRDs over 100 m in China

No	Engineering	Dam height / m	φ_0 /°	$\Delta\varphi$ /°	No	Engineering	Dam height / m	φ_0 /°	$\Delta\varphi$ /°
1	Shuibuya	233	51.2	9.1	13	Longma	135	51.7	11.0
			52.0	8.5	14	Shanxi	132	56.1	11.6
			50.0	8.4				54.4	9.8
			52.0	8.5	15	Yinzidu	130	50.6	10.9
2	Houziyan	223	49.6	7.5	16	Jiemian	126	52.5	9.5
			49.8	7.2				53.0	9.0
			48.0	7.5				52.0	10.0
			50.0	8.2	17	Eping	125	47.0	6.8
3	Jiangpinghe	210	47.7	6.3	18	Heiquan	124	47.0	6.8
			53.0	6.5				48.0	7.0
			48.6	5.5				46.0	6.5
			48.3	5.2	19	Baixi	124	47.3	6.3
4	Sanbanxi	186	51.6	8.5				47.0	7.0
			56.2	12.5				47.9	5.0
			55.7	12.4	20	Baiyun	120	53.2	6.4
			52.7	10.6	21	Qinshan	120	46.3	8.6
			55.7	12.4				50.0	6.1
			46.3	7.4				48.0	10.0
5	Hongjiadu	180	53.0	9.0	22	Gudongkou	118	48.4	7.2
			52.0	10.0	23	Sujiahekou	117	51.0	10.5
			52.3	7.3	24	Bajiaohe	115	49.9	10.0
			51.3	6.9				49.5	10.0
			57.0	13.1	25	Sinanjiang	115	42.8	8.3
			52.8	9.6	26	Gaotang	112	45.6	5.9
6	Tianshengqiao	178	52.9	11.5	27	Chahanwusu	110	52.8	10.0
			58.0	14.0				53.2	10.0
			51.6	16.7				51.4	9.3
			58.9	13.1				54.4	10.7
			54.0	13.0				53.2	10.4
			54.0	13.5				51.4	9.3
			48.0	10.0				54.4	10.7
			46.9	4.5	28	Nalan	109	49.4	7.3
			48.7	5.7				56.0	14.0
			53.5	8.7				47.0	6.4
			57.0	13.0	29	Yutiao	106	44.6	5.1

(continued)

Table 7.4 (continued)

No	Engineering	Dam height / m	φ_0 /°	$\Delta\varphi$ /°	No	Engineering	Dam height / m	φ_0 /°	$\Delta\varphi$ /°
			50.5	10.0				42.6	5.7
7	Tankeng	161	54.7	11.4				50.0	8.4
			55.8	12.0				42.0	5.5
			55.4	11.9	30	Qiezishan	106	48.6	8.3
			54.4	11.5	31	Liyutang	105	45.5	8.5
			56.0	12.1				47.0	8.0
			55.8	12.1				44.0	9.0
			54.7	11.5	32	Dongba	105	52.1	9.6
			55.4	11.9	33	Panshitou	103	54.2	12.4
			56.3	12.4				43.9	10.0
			57.1	12.4				44.9	11.5
8	Jilintai	157	55.9	12.6	34	Sianjiang	103	46.7	6.6
			49.8	9.2	35	Chaishitan	103	47.4	10.2
			51.2	13.5	36	Baishuikeng	101	52.6	14.3
			57.0	14.0				58.4	13.4
			53.0	8.1	37	Kajiwa	171	51.4	9.0
9	Zipingpu	156	53.6	11.2				50.6	8.7
			55.4	10.6				52.4	9.5
10	Malutang	154	51.9	10.6				52.6	9.5
			55.0	15.0	38	Cihaxia	254	54.2	8.1
11	Gongboxia	139	49.8	9.4				52.7	8.2
			50.0	9.0	39	Gushui	242	55.5	11.3
			46.7	8.1				53.0	11.0
			47.2	7.1				53.5	10.7
			45.2	13.0				54.4	10.6
			46.8	4.4				55.0	12.2
			52.0	9.8				52.0	9.9
			51.3	14.8	40	Jiudianxia	137	50.9	8.5
12	Wuluwati	138	43.9	3.3					
			43.5	3.0					
			44.2	3.6					

φ_0: Mean of 51.1; Std.D of 3.8; $\Delta\varphi$: Mean of 9.5; Std.D of 2.7

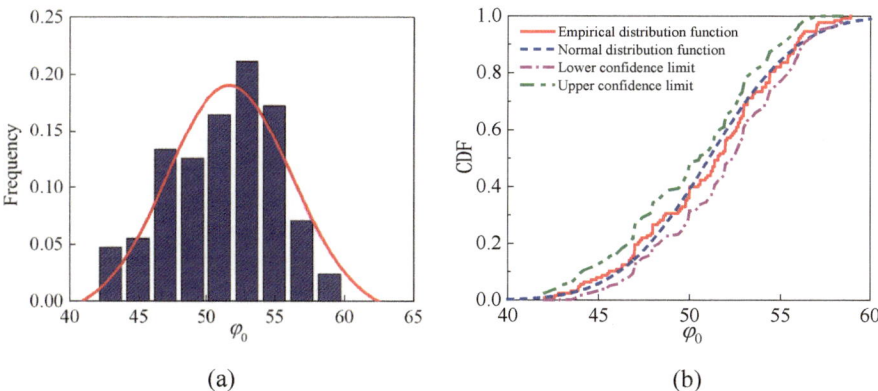

Fig. 7.13 Frequency histograms and statistical features of φ_0 **a** Frequency histogram **b** Comparison of CDFs

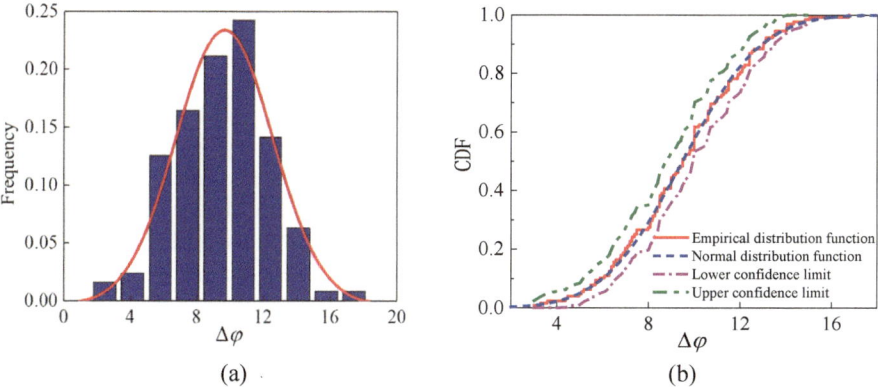

Fig. 7.14 Frequency histograms and statistical features of $\Delta\varphi$. **a** Frequency histogram. **b** Comparison of CDFs

Post-peak shear strain	$\varphi_0(°)$	$\Delta\varphi(°)$
0	51.1	9.5
0.03	50.1	8.8
0.06	49.2	8.1
0.09	48.2	7.5
0.12	47.2	6.9
0.15	46.3	6.2
0.18	45.3	5.6
0.21	44.3	4.9

Table 7.5 The post-peak shear strain and strength based on the mean of shear strength

7.5 Dam Slope Stability Performance Evaluation Considering Randomness of Ground Motion

In this section, the randomness of the ground motion is considered. The GPDEM, reliability probability analysis method and fragility analysis method are employed to assess the seismic safety of a 250-m CFRD slope considering the softening effects of the rockfill materials. This evaluation reveals the evolution patterns of its stochastic dynamic response and changes in failure probability under different levels of seismic intensity, providing valuable references for the safety evaluation of dam slope stability.

7.5.1 Basic Information

The calculation models, load conditions remain consistent with those described in Sect. 3.4.1. A Duncan-Chang E-B constitutive model was employed for the static calculations and then the initial stress conditions were provided for the subsequent dynamic calculations. The Hardin-Drnevich constitutive model was adopted to simulate the non-linear behaviour of the rockfill materials during the dynamic calculations. The rockfill material parameters of the two constitutive models are provided in Tables 7.6 and 7.7, respectively. The parameters of bedrock and faced-slabs are the same as in Sect. 3.4.1; The peak and post-peak strengths are listed in Table 7.5 in order to consider the softening effects. The PGA was adjusted from 0. 1 to 1.0 g with 0. 1 g intervals, resulting in a total of 1440 acceleration time histories.

Table 7.6 Parameters for duncan E-B model

ρ /(kg/m3)	K	n	R_f	K_b	m	φ_0 /(°)	$\Delta\varphi$ /(°)
2160	1350	0.28	0.80	780	0.18	51.1	9.5

Table 7.7 Parameters for Hardin-Drnevich model

K	n	ν
2660	0.444	0.33

7.5.2 Stochastic Dynamic Response for Slope Stability of High CFRDs

(1) safety factor

Figure 7.15 illustrates the safety factor time histories of three typical samples and the mean and standard deviation time histories with PGA = 0. 1, 0. 5 and 1. 0 g, respectively. As can be seen, when the seismic intensity increases, the safety factor exhibits more drastic fluctuations over time, and the standard deviation also increases. The mean time history of safety factor tends to be stable, indicating that the random ground motion samples generated have good statistical characteristics. Figure 7.16 illustrates the probability information for safety factor derived by using GPDEM with PGA = 0. 5 g. The PDFs at three typical times are entirely distinct and display two or even multiple peaks, not fitting the regular distributions like normal or log-normal distributions. The PDF evolution surface vividly illustrates the evolving process of the PDF of safety factor over time which has significant fluctuations and transformations, as shown in Fig. 7.16b. The evolution of the PDF also indicates larger variability among different seismic samples, and the fluctuations and evolutions of this variability is more evident in the PDF contour map, as shown in Fig. 7.16c.

Figure 7.17 presents the discrete point distribution of minimum safety factor with PGA = 0. 5 g and 1. 0 g. The point distribution has a large dispersion, and the

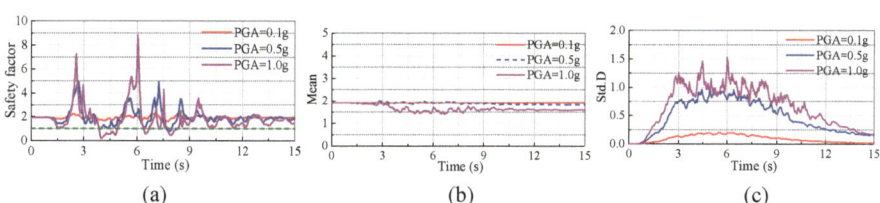

Fig. 7.15 Response, mean and standard deviation time history of safety factor **a** Safety factor **b** Mean **c** Standard deviation

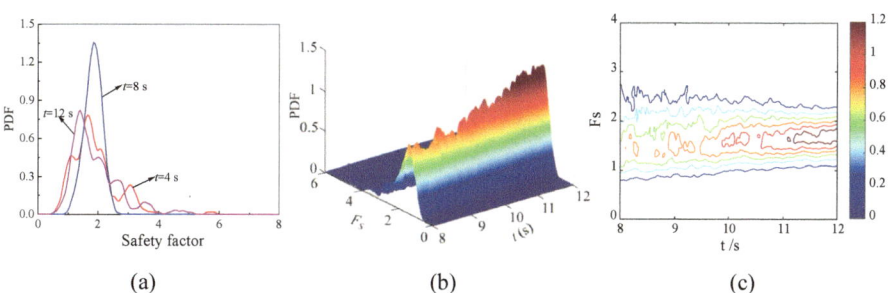

Fig. 7.16 Probability evolution information of safety factor.**a** PDFs at three typical times. **b** PDF evolution surface. **c** PDF contour map

difference between the maximum and minimum values is significant. The maximum value with PGA $= 0.5$ g is 1.17 and the minimum value is 0.42, while the maximum value with PGA $= 1.0$ g is 0.66 and the minimum value is 0, which indicates that the safety factor has high statistical significance subjected to stochastic earthquake, emphasizing the importance of analyzing from the perspective of stochastic dynamics based on the minimum safety factor. The exceedance probability increases at a faster rate with higher seismic intensity, seen from Fig. 7.18. This may be attributed to the softening effects of rockfill materials under strong earthquake. And the minimum safety factors under different seismic intensities are primarily distributed between the 95% and 5% exceedance probabilities. The numerical range of safety factor is as follows: 1.31–1.87, 1.05–1.64, 0.84–1.45, 0.67–1.30, 0.50–1.17, 0.35–1.04, 0.19–0.92, 0.08–0.80, 0–0.71, and 0–0.61. These findings serve as a reference for the seismic stability design of high CRFD under earthquake.

(2) Cumulative time of $F_S < 1.0$

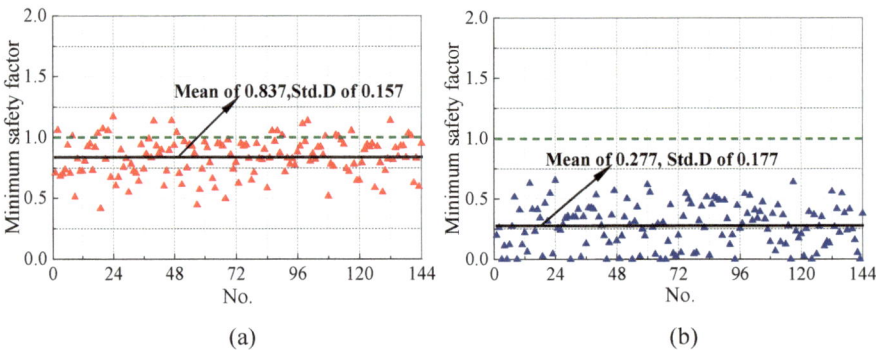

Fig. 7.17 Discrete point distribution of minimum safety factor under 0.5 g and 1.0 g. **a.** PGA $= 0.5$ g .**b** PGA $= 1.0$ g

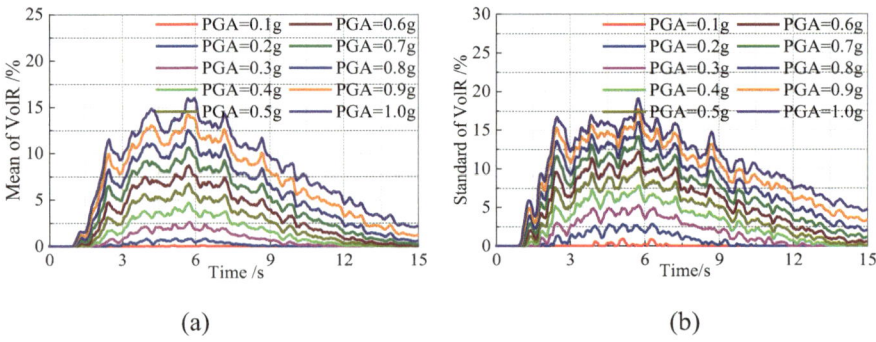

Fig. 7.18 Exceedance probability of minimum safety factor under different PGA. **a** Exceedance probability. **b** The minimum safety factor

Figure 7.19 presents the discrete point distribution of cumulative time of $F_S < 1.0$ with PGA = 0.6 g and 1. 0 g. The point distribution has a large dispersion, and the difference between the maximum and minimum values is significant. The maximum value with PGA = 0.6 g is 2.31 s and the minimum value is 0 s, while the maximum value with PGA = 1. 0 g is 4.02 s and the minimum value is 0.65 s. This indicates the importance of analyzing from the perspective of stochastic dynamics based on the cumulative time with $F_S < 1.0$. Figure 7.20 shows the exceedance probability of cumulative time with $F_S < 1.0$ under different seismic intensities. The probabilities primarily range between 5 and 95% and the numerical ranges for each seismic intensity are as follows: 0–0.11 s (0.3 g), 0–0.48 s (0.4 g), 0–1.27 s (0. 5 g), 0.02–1.85 s (0.6 g), 0.21–2.30 s (0.7 g), 0.51–2.71 s (0.8 g), 0.77–3.12 s (0.9 g), and 1.18–3.51 s (1. 0 g). These findings provide references for performance-based seismic stability evaluation of high CRFD.

(3) Cumulative slippage

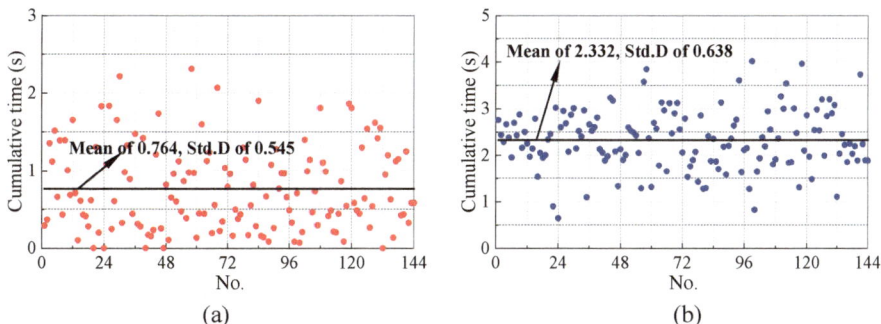

Fig. 7.19 Distribution of cumulative time with $F_S < 1.0$ under 0.6 g and 1. 0 g. **a** PGA = 0.6 g. **b** PGA = 1. 0 g

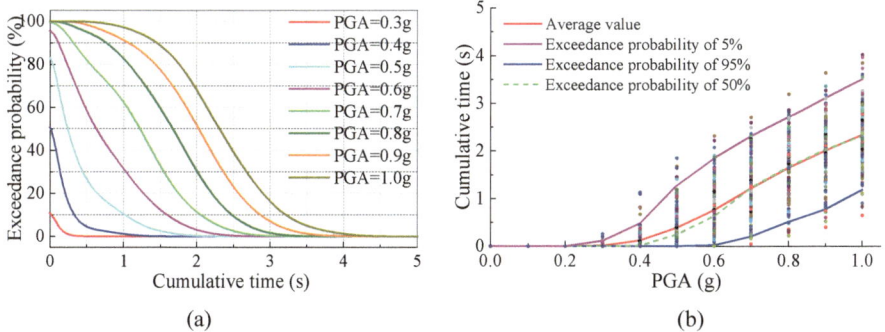

Fig. 7.20 Exceedance probability of cumulative time with $F_S < 1.0$ under different PGA. **a** Exceedance probability. **b** The cumulative time with $F_S < 1.0$

Figure 7.21 illustrates the slip surfaces corresponding to the minimum safety factor with PGA $= 0.4$ g and 0.6 g. It is evident that the position and size of the slip surfaces vary with different seismic intensities. The slip surface is smaller but more $+$ prone to shallow sliding under strong earthquakes. Figure 7.22 illustrates the discrete point distribution of the cumulative slippage with PGA $= 0.6$ g and 1. 0 g. The point distribution has a large dispersion, and the difference between the maximum and minimum values is very significant. The maximum value with PGA $= 0.6$ g is 331 cm and the minimum value is 0 cm, while the maximum value with PGA $= 1. 0$ g is 848 cm and the minimum value is 13 cm and this indicates the importance of analyzing from the perspective of stochastic dynamics based on the cumulative slippage. Figures 7.23 and 7.24 shows the exceedance probability and scatter plot of cumulative slippage under different seismic intensities. The probabilities primarily range between 5 and 95%. The numerical ranges for each seismic intensity are as follows: 0–0.3 cm (0.3 g), 0–11 cm (0.4 g), 0–67 cm (0. 5 g), 0–148 cm (0.6 g), 0–261 cm (0.7 g), 5–350 cm (0.8 g), 15–498 cm (0.9 g), 46–624 cm (1.0 g). These findings provide references for the performance-based stability assessment of high CRFD.

(4) Discussion on the relationship between cumulative time of $F_S < 1.0$ and cumulative slippage

The cumulative time of $F_S < 1.0$ and the cumulative slippage both exhibit cumulative effects, suggesting a certain degree of correlation. Therefore, this section investigates

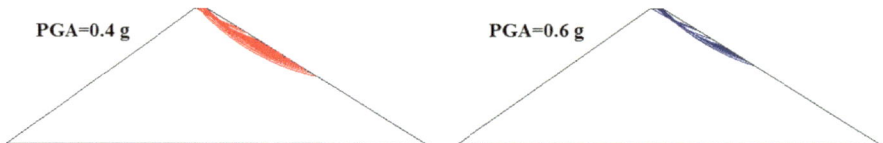

Fig. 7.21 The slide surface corresponding to the minimum safety factor

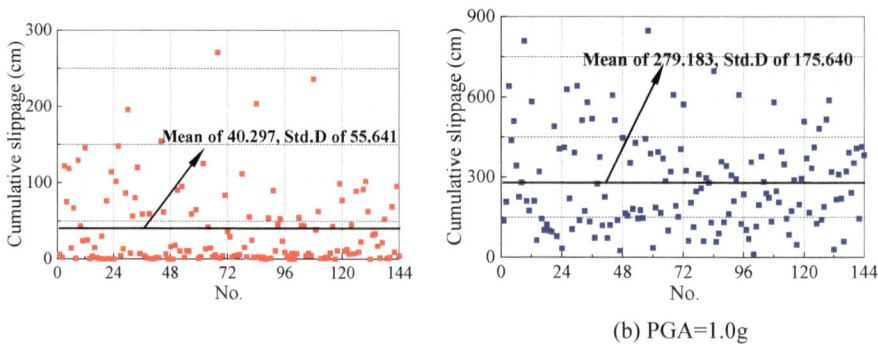

Fig. 7.22 Distribution of cumulative slippage under 0.6 g and 1. 0 g. **a** PGA $= 0.6$ g. **b** PGA $= 1. 0$ g

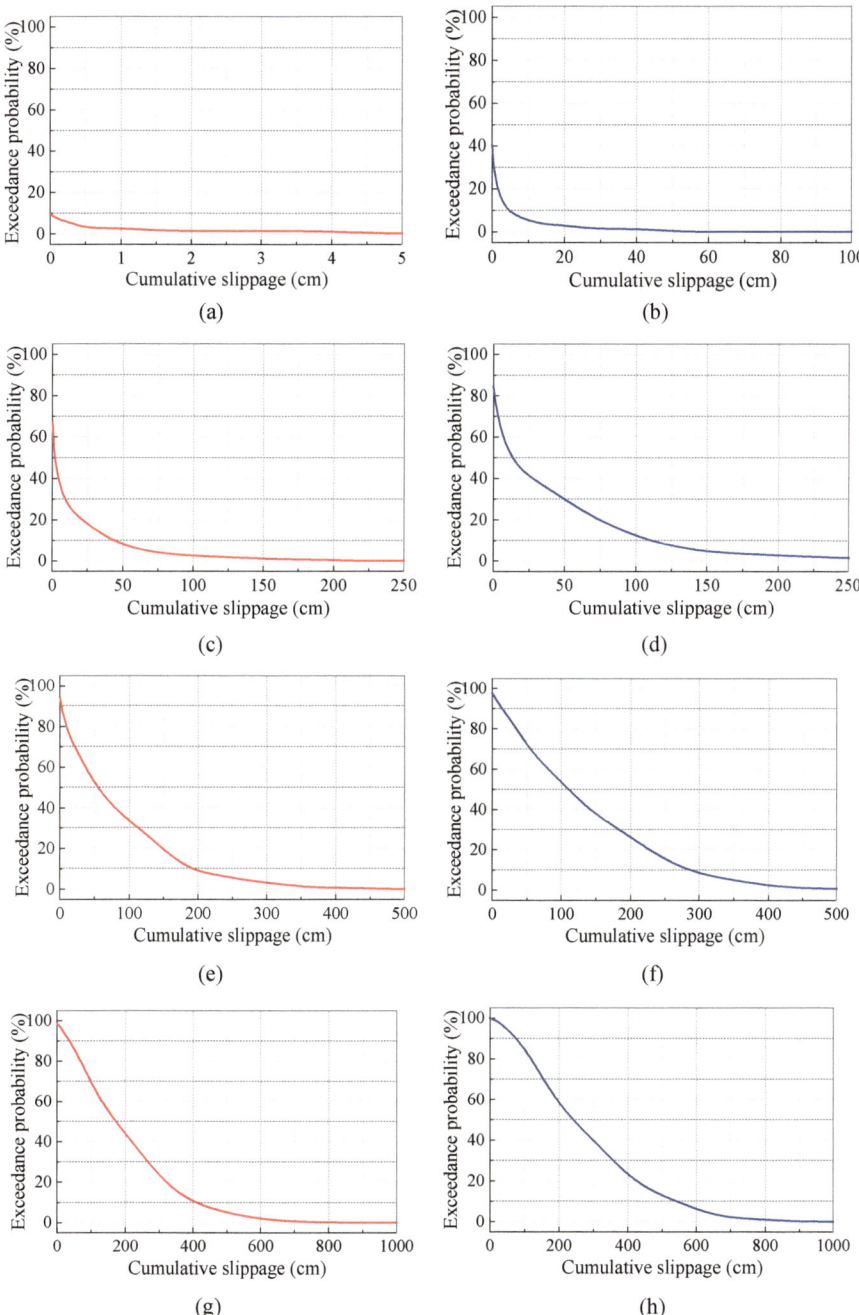

Fig. 7.23 Exceedance probability of cumulative slippage under different PGA. **a** PGA = 0.3 g. **b** PGA = 0.4 g. **c** PGA = 0.5 g. **d** PGA = 0.6 g. **e** PGA = 0.7 g. **f** PGA = 0.8 g. **g** PGA = 0.9 g. **h** PGA = 1.0 g

Fig. 7.24 Cumulative
slippage under different PGA

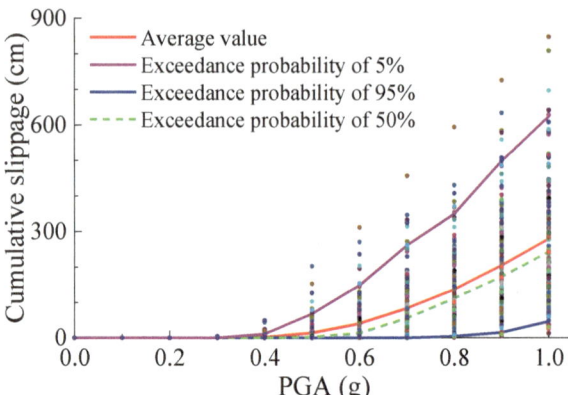

the preliminary relationship between the cumulative time of $F_S < 1.0$ and the cumulative slippage under different seismic intensities through extensive sample analysis. It can be observed that there is a certain correlation between the cumulative time of $F_S < 1.0$ and the cumulative slippage, seen from Fig. 7.25. However, the correlation decreases as the seismic intensity increases. Based on the distribution pattern, it is inferred that a complex correlation exists between the two indices and it is influenced by seismic intensity.

7.6 Dam Slope Stability Stochastic Dynamic Analysis Considering Randomness of Shear Strength Parameters

7.6.1 Basic Information

The working conditions and load conditions of the high CFRD are the same as those in Sect. 3.4.1. The static and dynamic parameters are adopted those in Sect. 7.5.1 The mean and standard deviation of φ_0 and $\Delta\varphi$ as well as the type of distribution are the values obtained in Sect. 7.4, and 144 sets of shear strength parameters were obtained based on the GF-discrepancy method. Considering the softening effects, the post-peak strengths of φ_0 and $\Delta\varphi$ is obtained by using Eqs. (7.10) and (7.11), respectively. A series of finite element dynamic stability calculations were performed with PGA $= 0.5$ g, and the stochastic dynamic and probabilistic information of the safety factor, the cumulative time of $F_S < 1.0$ and the cumulative slippage were obtained by combining with the GPDEM, so as to provide a reference for safety evaluation of dam slope stability performance based on parameter randomness.

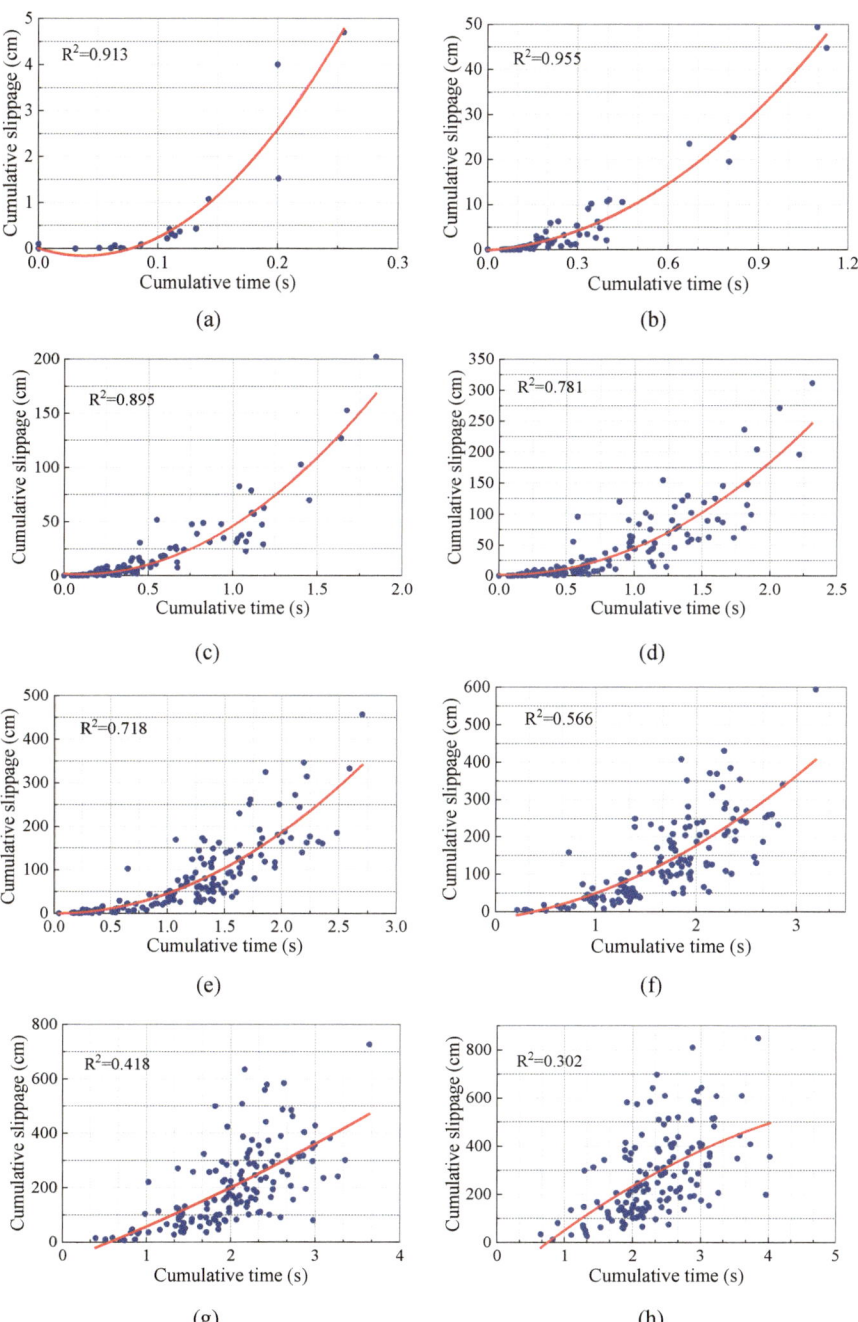

Fig. 7.25 The relationship between cumulative time and cumulative slippage. **a** PGA = 0.3 g. **b** PGA = 0.4 g. **c** PGA = 0. 5 g. **d** PGA = 0.6 g. **e** PGA = 0.7 g. **f** PGA = 0.8 g. **g** PGA = 0.9 g. **h** PGA = 1. 0 g

7.6.2   Safety Factor

Figure 7.26 illustrates the safety factor time history with deterministic mean material parameters and the mean and standard deviation time history of safety factor. It can be observed that the safety factor exhibits different patterns compared to seismic randomness, indicating the significance of considering the material parameters uncertainties. The significant fluctuations in the standard deviation further emphasize the strong influence of material parameters uncertainties on the safety factor under seismic excitation.

Figure 7.27 shows that different material parameters lead to varying minimum safety factors, with a range between 1.19 and 0.46. The value at 95% exceedance probability is 0.48, while it is 1.10 at 5% exceedance probability, with a difference of approximately 2.3 times. There are noticeable distinctions between the safety factor exceedance probabilities considering the seismic randomness and material parameter randomness. Hence, it is essential to consider the effects of material parameters randomness on safety factor.

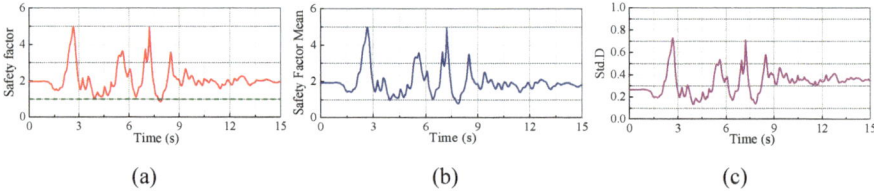

(a)   (b)   (c)

Fig. 7.26 Response, mean and standard deviation time history of safety factor. **a** Safety factor. **b** Mean. **c** Standard deviation

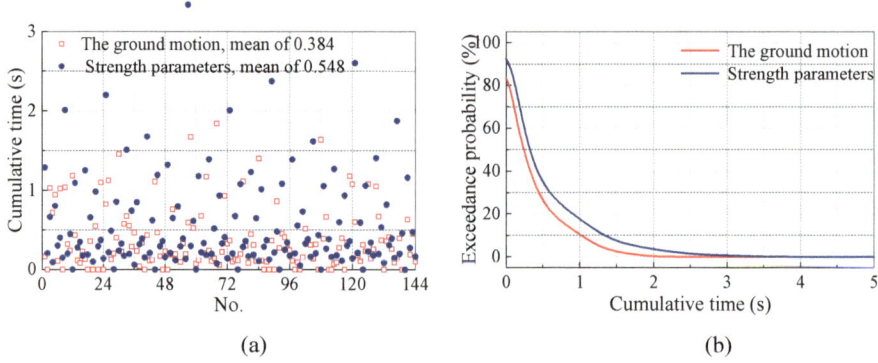

(a)   (b)

Fig. 7.27 Distribution and exceedance probability of minimum safety factor. **a** Distribution. **b** Exceedance probability

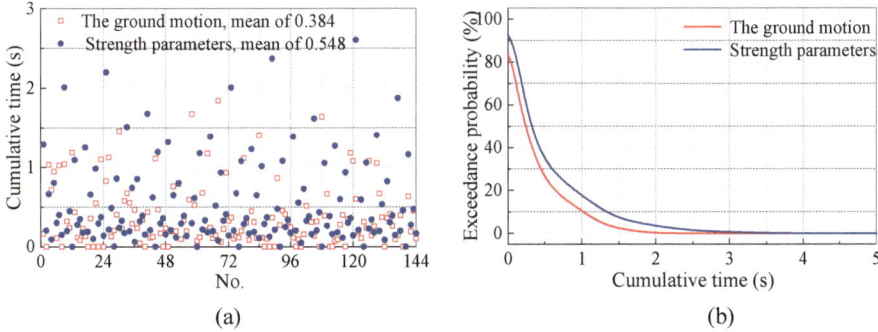

Fig. 7.28 Distribution and exceedance probability of cumulative time. **a** Distribution. **b** Exceedance probability

7.6.3 Cumulative Time of $F_S < 1.0$

Figure 7.28 shows that different material parameters lead to varying cumulative time of $F_S < 1.0$, with a range between 3.34 and 0 s. The value at 95% exceedance probability is 0 s, while it is 1.77 s at 5% exceedance probability. There are noticeable distinctions between the safety factor exceedance probabilities considering the seismic randomness and material parameter randomness. Hence, it is essential to consider the impacts of the material parameters randomness on the cumulative time of $F_S < 1.0$. According to the damage level division criteria in Sect. 8, the corresponding probabilities for mild damage (0 s), moderate damage (0.5 s), and severe damage (1.5 s) are 92.4%, 35.3%, and 7.6%, respectively.

7.6.4 Cumulative Slippage

Figure 7.29 shows slip surfaces corresponding the minimum safety factor obtained considering seismic randomness (red line) and shear strength parameters randomness (blue line) with PGA = 0. 5 g. It is evident that different randomness significantly affects the position and size of the slip surfaces. Figure 7.30 shows that different material parameters result in varying cumulative slippage, with a significant difference between the maximum value of 363 cm and the minimum value of 0 cm. The value at 95% exceedance probability is 0 cm, while it is 141 cm at 5% exceedance probability, indicating a substantial variation. There are noticeable distinctions between the cumulative slippage considering the seismic randomness and material parameter randomness. Hence, it is crucial to consider the effects of the material parameters randomness on the cumulative slippage. According to the damage level division criteria in Sect. 8, the corresponding probabilities for mild damage (0 cm), moderate damage (20 cm), and severe damage (100 cm) are 80.1%, 28.9%, and 7.1%, respectively. The correlation analysis of cumulative time and cumulative slippage shown

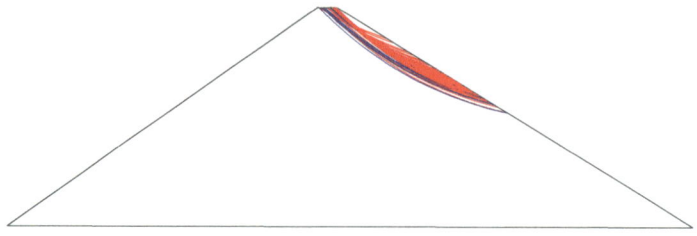

Fig. 7.29 The slide surface corresponding to the minimum safety factor

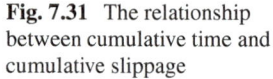
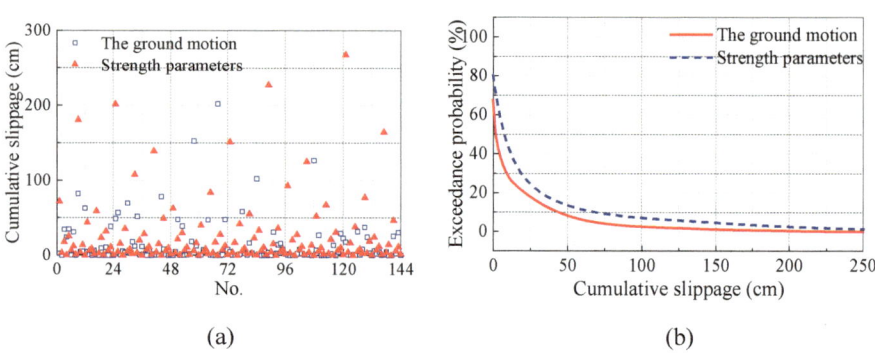

(a)　　　　　　　　　　　　　　(b)

Fig. 7.30 Distribution and exceedance probability of cumulative slippage. **a** Distribution. **b** Exceedance probability

Fig. 7.31 The relationship between cumulative time and cumulative slippage

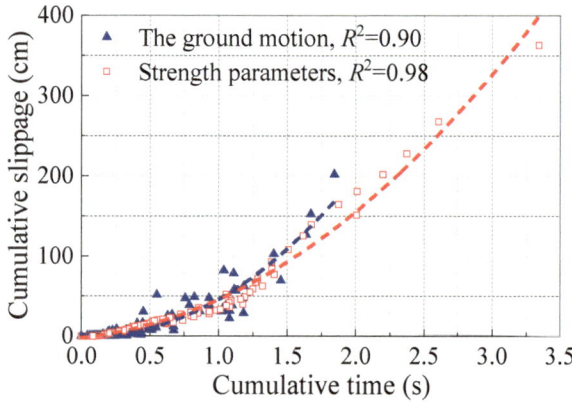

in Fig. 7.31 reveals that considering material parameter randomness results in better correlation between the two factors.

References

Chen Z, Morgenstern NR, Chan DH (1992) Progressive failure of the Carsington Dam: a numerical study. Can Geotech J 29(6):971–988

Guan Z (2009) Investigation of the 5.12 wenchuan earthquake damages to the zipingpu water control project and an assessment of its safety state. Sci China Ser E: Technol Sci 52:820–834

Liu H, Ling HI (2012) Seismic responses of reinforced soil retaining walls and the strain softening of backfill soils. Int J Geomech 12(4):351–356

Liu H, Chen Y, Yu T, Yang G (2015) Seismic analysis of the Zipingpu concrete-faced rockfill dam response to the 2008 Wenchuan, China, earthquake. J Perform Constr Facil 29(5):04014129

Liu J, Liu F, Kong X, Yu L (2016) Large-scale shaking table model tests on seismically induced failure of concrete-faced rockfill dams. Soil Dyn Earthq Eng 82:11–23

Potts DM, Dounias GT, Vaughan PR (1990) Finite element analysis of progressive failure of Carsington embankment. Géotechnique 40(1):79–101

Skempton AW (1985) Residual strength of clays in landslides, folded strata and the laboratory. Geotechnique 35(1):3–18

Uddin N (1999) A dynamic analysis procedure for concrete-faced rockfill dams subjected to strong seismic excitation. Comput Struct 72(1–3):409–421

Wang XR, Rong QG, Sun SL, Wang H (2017) Stability analysis of slope in strain-softening soils using local arc-length solution scheme. J Mt Sci 14:175–187

Zhang B, Li D (2014) Review on study of ultimate aseismic capacity of high dams. Water Res Power 32(1):63–65

Zhang G, Zhang JM (2007) Simplified method of stability evaluation for strain-softening slopes. Mech Res Commun 34(5–6):444–450

Zhu Y, Kong X, Zou D, Li Y (2011) Shaking table test and numerical simulation on the effect of reinforcement on the seismic safety of dam slopes. J Harbin Inst Technol 18(4):132–138

Zhu Y, Zou D, Xu B, Teng X (2016) Dynamic stability and slip deformation of dam slope considering softening of rockfill materials. Chinese J Geotech Eng 38(9):1713–1719

Zou D, Xu B, Kong X, Liu H, Zhou Y (2013) Numerical simulation of the seismic response of the Zipingpu concrete face rockfill dam during the Wenchuan earthquake based on a generalized plasticity model. Comput Geotech 49:111–122

Chapter 8
Performance Seismic Safety Evaluation

8.1 Seismic Safety Evaluation of CFRD Considering the Randomness of Ground Motion

CFRDs are inherently complex, and their dynamic response and seismic damage under seismic action are manifested in various aspects. However, due to the scarcity of seismic damage data and seismic codes specifically tailored for earth and rock dams, the focus remains primarily on three aspects: deformation of the dam body, stability of the dam slopes, and safety of the panels of the seepage control body. It is believed that the deformation of the dam body may influence the overall performance of the dam, while the stability of the dam slopes and the safety of the panels may affect local functionality to some extent. In this section, drawing upon the extensive finite element dynamic calculations and stochastic dynamic response analyses discussed earlier, we utilize the generalized probability density evolution method to establish the relationship between multi-seismic intensity, multiple performance targets, and destruction probability. Initially, we focus on the deformation of the dam body and the safety of the impermeable panel body. Based on corresponding performance indexes, we propose a classification standard for performance levels. Subsequently, susceptibility probability analyses are conducted for various performance indexes under different damage levels, aiming to develop a performance-based framework for evaluating the seismic safety of CFRDs under the stochastic effects of ground shaking.

8.1.1 Dam Deformation

Liu et al. (2012) conducted a statistical analysis of the seismic settlement rates of 123 earth-rock dams with heights exceeding 15 m worldwide. Combining

B. Xu and R. Pang, *Stochastic Dynamic Response Analysis and Performance-Based Seismic Safety Evaluation for High Concrete Faced Rockfill Dams*, Hydroscience and Engineering, https://doi.org/10.1007/978-981-97-7198-1_8

publicly available seismic damage data and numerical calculation results, the numerical analysis mainly referred to high earth-rock dam projects under construction or planned in China, including 11 dams such as Nuozhadu, Lianghekou, Shuangjiangkou, Longpan, Houziyan, Liangfengtai, Wenquan, Jishi Gorge, and Longshou. It concluded that the majority of dam crest relative settlement rates are below 1%, with those exceeding 1% being primarily earth or hydraulic fill dams. It is preliminarily indicated that earth-rock dams constructed using modern heavy compaction techniques experience a dam crest relative settlement rate of approximately 1% under seismic peak ground acceleration less than 0.6 g. The Nanjing Hydraulic Research Institute (1998) initially suggested allowing a settlement rate of 2% for earth-rock dams below 100 m, 1.5% for those above 100 m, and 1.0% for those above 200 m. Zhao et al. (2015) proposed setting the dam crest relative settlement rate between 0.6% and 0.8% as the ultimate control standard for high panel rockfill dams. Tian et al. (2013) recommended setting the evaluation limits for dam heights of 100 m, 150 m, 200 m, 250 m, and 300 m at 2%, 1.5%, 1%, 0.85%, and 0.75% of the dam crest relative settlement rate, respectively. Chen et al. (2013) proposed using a dam crest settlement rate below 1% as the seismic deformation control criterion for core wall rockfill dams and below 0.6% as the ultimate control standard for high panel rockfill dams. Swaisgood et al. (2003) studied a total of 69 earth-rock dams (including panel rockfill dams, core wall rockfill dams, hydraulic fill dams, and earth dams) both domestically and internationally, using dam crest relative settlement rate as an indicator. They categorized the damage situation into intact (below 0.1%), slight damage (0.012%-0.5%), moderate damage (0.1%-1.0%), and severe damage (above 0.5%), but primarily focused on lower earth-rock dams. Figure 8.1 shows the curves for the relationship between the average PGA-dam top relative seismic subsidence rate and the 5% exceedance probability. It can be seen that the relative subsidence at the top of the dam increases with the increase in PGA, but the trend change gradually becomes slower. Combined with the literature in above, 0.3%, 0.7%, and 1.0% of the relative subsidence are suggested as the criticality of the performance level, which corresponds to the critical states of mild, moderate, and severe damages. In the Wenchuan earthquake, the 156 m-high Zipingpu faced rockfill dam suffered PGA = 0.55 g earthquake, and the subsidence was 0.81 m, which was about 0.519% of the height of the dam (Zhou 2012). According to the average fitting formula, the relative subsidence rate of the dam top was 0.521%, which was almost no different from the monitoring value, which proved the accuracy of the stochastic dynamic analysis, and it was in the range between mild and moderate damage according to the damage classification standard, consistent with the idea of 'medium damage can be repaired. According to the damage classification criteria, the dam is in the range of mild-moderate damage, which is consistent with the idea of 'medium earthquake can be repaired', and the Zipingpu dam has been repaired and is in normal use. In addition, we suggest that the derivative of the curve obtained from the 5% beyond probability fit is 0, which corresponds to the no-breakage critical state, at which time, PGA = 1.207 g, corresponding to a relative seismic subsidence of the dam top of 1.151%, which is taken as 1.1% for safety.

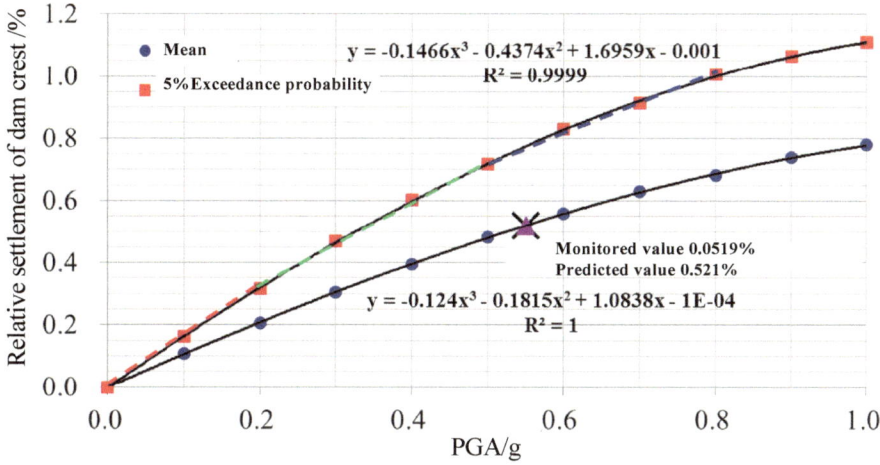

Fig. 8.1 Relationship curve of PGA-Relative subsidence of dam crest

Table 8.1 lists the exceedance probabilities corresponding to different levels of dam crest relative settlement under various seismic intensities. This establishes a relationship table between multiple seismic intensities, dam crest relative settlement, and exceedance probability, providing reference for performance-based seismic safety design. The significance lies in determining control standards for dam crest relative settlement corresponding to different exceedance probabilities, based on the owner's or practical requirements. According to Table 8.1, the exceedance probability for the Zipingpu CFRD faced rockfill dam falls between 50 and 60%, indicating a relatively high probability of seismic damage occurrence, which aligns with the actual situation. Furthermore, Table 8.1 holds significant reference value for studying the ultimate seismic resistance capacity of high CFRDs. The following recommendations are proposed: For PGA values ranging from 0.1 to 0.3 g, an exceedance probability of approximately 25% is suggested. At PGA = 0.1 g, the relative settlement is 0.15%; at PGA = 0.2 g, it is 0.25%; and at PGA = 0.3 g, it is 0.35%. For PGA values between 0.4 and 0.7 g, an exceedance probability of around 50% is recommended. Specifically, at PGA = 0.4 g, the relative settlement is 0.4%; at PGA = 0.5 g, it is 0.5%; at PGA = 0.6 g, it is 0.55%; and at PGA = 0.7 g, it is 0.6%. For PGA values between 0.8 and 1.0 g, an exceedance probability of approximately 75% is suggested. At PGA = 0.8 g, the relative settlement is 0.55%; at PGA = 0.9 g, it is 0.6%; and at PGA = 1.0 g, it is 0.65%. Figure 8.2 depicts the damage probability corresponding to different peak accelerations for various damage levels, i.e., fragility curves, obtained through B-Spline interpolation fitting. Different performance levels of damage probability can be defined based on these curves.

Furthermore, in line with the design philosophy of "minor earthquakes do not cause damage, moderate earthquakes are reparable, and major earthquakes do not cause collapse," the following recommendations are proposed: For minor earthquakes (0–0.2 g), the dam remains essentially intact, with only a very small percentage

Table 8.1 Relationship of multiple earthquake intensities-relative settlement rate of dam crest-exceedance probability

Exceedance probability/%		Relative seismic subsidence rate at dam roof/%											
		0.1	0.2	0.3	0.4	0.5	0.6	0.7	0.8	0.9	1.0	1.1	1.2
PGA	0.1 g	55.9	0	0	0	0	0	0	0	0	0	0	0
	0.2 g	97.0	51.5	7.4	0	0	0	0	0	0	0	0	0
	0.3 g	99.3	86.5	49.2	15.3	2.8	0	0	0	0	0	0	0
	0.4 g	100	96.1	77.8	45.9	18.6	5.2	0	0	0	0	0	0
	0.5 g	100	98.9	91.5	70.5	41.9	18.8	6.2	0	0	0	0	0
	0.6 g	100	100	96.1	84.1	62.1	37.8	18.5	6.7	0	0	0	0
	0.7 g	100	100	98.3	91.9	76.9	54.8	32.5	15.4	5.7	0	0	0
	0.8 g	100	100	100	93.3	82.1	65.2	44.8	26.4	13.1	5.4	1.8	0
	0.9 g	100	100	100	96.5	88.5	75.1	55.4	35.9	19.8	9.3	3.5	1.1
	1.0 g	100	100	100	97.8	92.6	81.3	64.1	44.1	26.1	13.1	5.4	1.9

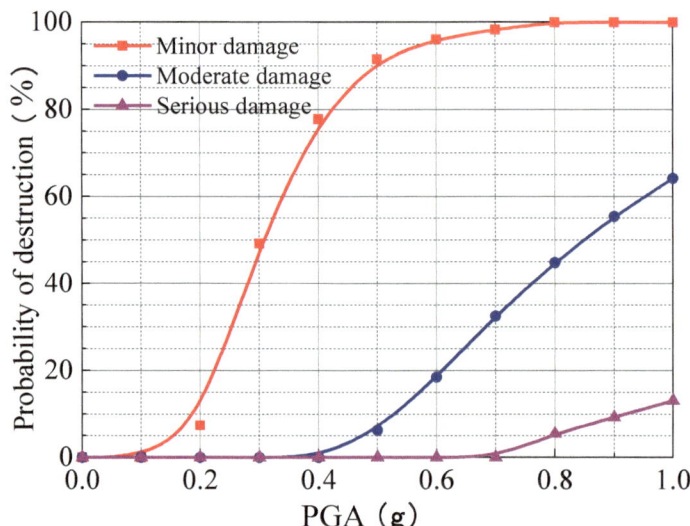

Fig. 8.2 Fragility curve based on relative settlement rate of dam crest

(below 5%) experiencing "mild damage." Instances of mild damage or above are rare, and the relative settlement rate at the dam crest is 0.3% when subjected to seismic action of 0.2 g, corresponding to a damage probability of around 7.4%, approximately 5%. For moderate earthquakes (0.2–0.5 g), the dam is in a state of "mild to moderate" damage, with only a very small percentage (below 5%) reaching moderate damage. Instances of moderate damage or above are rare, and the relative settlement rate at the dam crest is 0.7% when subjected to seismic action of

0.5 g, corresponding to a damage probability of around 6.2%, approximately 5%. For major earthquakes (0.5–0.8 g), the dam is in a state of "moderate to severe" damage, with only a very small percentage (below 5%) reaching severe damage. Instances of dam collapse are rare, and the relative settlement rate at the dam crest is 1.0% when subjected to seismic action of 0.8 g, corresponding to a damage probability of around 5.4%, approximately 5%. For earthquakes exceeding 1.0 g, a very small percentage (below 5%) of dam collapses may occur. The exceedance probability of a dam crest settlement rate of 1.1% under seismic action of 1.0 g is around 5.4%, approximately 5%, and earthquakes exceeding 1.0 g are generally rare. Therefore, the above analysis from a probabilistic perspective essentially demonstrates the rationality of the performance level classification criteria mentioned above.

8.1.2 Panel Impermeable Body Safety

In the evaluation of seismic safety of panel impermeable body, the current related research mainly focuses on panel stress and joint deformation as the control criteria, however, the short time panel overstress may not cause damage, so the effect of overstress holding time should be considered. Ghanaat (2004) used the results of elastic time-range analysis to propose the damage classification standard of concrete dams based on the ratio of tensile stress to concrete tensile strength obtained by calculation, i.e., the demand capacity ratio (DCR) and the cumulative overstress duration (COD) that the dam tensile stress is greater than the concrete tensile strength in the seismic process, i.e., the accumulated time of overstress holding time, and the damage classification standard of concrete dams was proposed, and the COD of 0.4 s was determined when the DCR = 1 and 0 s for the DCR = 2, and the damage limit of slight and moderate damage was determined. When DCR = 1, the COD is 0.4 s, and when DCR = 2, the COD is 0 s, which determines the boundaries of slight and moderate damage, and this concept was introduced in the seismic performance evaluation of concrete gravity dams by Shen et al (2007).

The DCR of a concrete panel can be defined as the ratio of the tensile stress of the panel to the tensile strength of the concrete:

$$DCR = \sigma_t/f_t. \tag{8.1}$$

where σ_t is the tensile stress for the linear elastic analysis of the panel; ft is the static tensile strength of the concrete of the panel, which can be obtained from the formula suggested by Raphael (1984):

$$f_t = 0.325(f_c)^{2/3} \tag{8.2}$$

f_t is the compressive strength of plain concrete. Since the dynamic tensile strength of concrete material is increased more than the static tensile strength due to the effect of seismic deformation rate and the assumption of linear elasticity finite element

Fig. 8.3 Static and dynamic nominal tensile strength for concrete

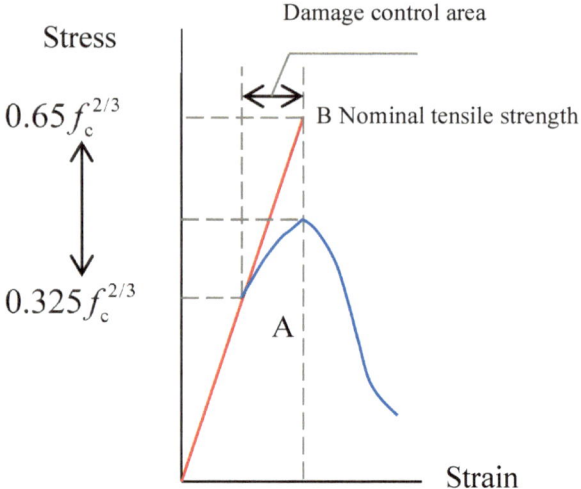

factor, Raphael suggests that the nominal tensile strength of concrete under seismic dynamic loading is:

$$f_d = 0.65(f_c)^{2/3} \tag{8.3}$$

Therefore, the maximum allowable DCR value for the panel concrete in the linear elasticity analysis is divided by the nominal tensile strength under seismic loading (DCR $= 2$), and the corresponding stress is two times the static tensile strength, as shown in Fig. 8.3.

Cumulative Overstress Duration (COD) is defined as the total duration of tensile stress when DCR ≥ 1. The longer the cumulative duration, the greater the possibility of panel damage. The permissible COD is determined as follows: the COD of five simple harmonic stresses when the stress harmonic amplitude with a period of T $= 0.2$ s reaches two times the static tensile strength (DCR $= 1$) is calculated by Eq. (8.4), as shown in Fig. 8.4. For DCR $= 1$, the COD is 0.6 s (recommended value for panel dams); when DCR $= 2$, the COD is 0 s, and Fig. 8.5 shows the time history of panel stress change under the action of two different intensities of ground shaking.

$$t_s \sum_{i=1}^{5} t_i \tag{8.4}$$

Therefore, according to the dynamic change of panel stress, the preliminary evaluation of panel safety and damage classification, the proposed damage levels are as follows: no damage or mild damage, the panel does not exceed the tensile strength, DCR ≤ 1, the panel is in the linear elasticity range; mild-moderate damage, the panel is cracked, but at an acceptable level, the DCR and the COD are in the shaded part of

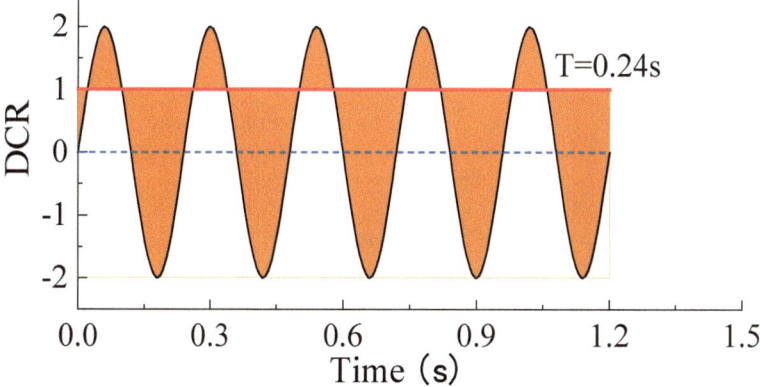

Fig. 8.4 Cumulative overstress duration (COD)

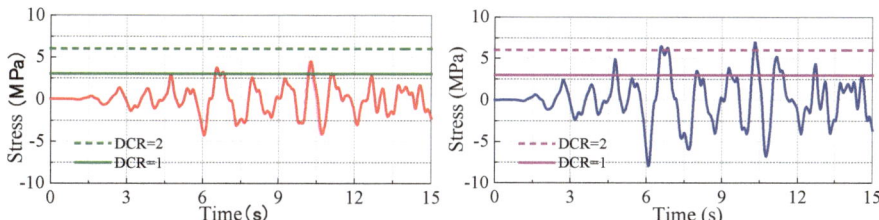

Fig. 8.5 Stress time history under earthquake action with different intensity

Fig. 8.6. Moderate-heavy damage or severe damage with DCR $\geq$ 2 or COD outside the shaded portion of Fig. 8.6 should be analyzed using nonlinear elastic–plastic time-course analysis to further assess the degree of damage and seismic safety of the panels.

Figure 8.7 shows the exceeding probability of COD under different DCR, e.g., DCR $=$ 1.5 indicates that the COD is greater than 1.5; when PGA $=$ 0.1 g, the COD $=$ 0 at DCR $=$ 1, which indicates that the stresses are all within the line elasticity range, and the panel is not damaged. And the exceeding probability that the panel damage indexes are in different ranges under each seismic intensity can be obtained, which provides the basis for performance-based seismic safety evaluation of CFRD from the perspective of panel damage.

From the above exceedance probability to get the fragility curves as shown in Fig. 8.8, when PGA $=$ 0.1 g, the panel is completely in the linear elastic range, no damage occurs; when PGA $=$ 0.2 g, the panel occurs mild damage probability of 28.6%, the probability of moderate damage 3.4%; when PGA $=$ 0.3 g, the panel occurs mild damage probability of 77.5%, the probability of moderate damage 37.1%; when PGA $=$ 0.4 g, the probability of mild destruction of the panel is 92.5% and the probability of moderate destruction is 69.9%; when PGA $=$ 0.5 g, the probability of

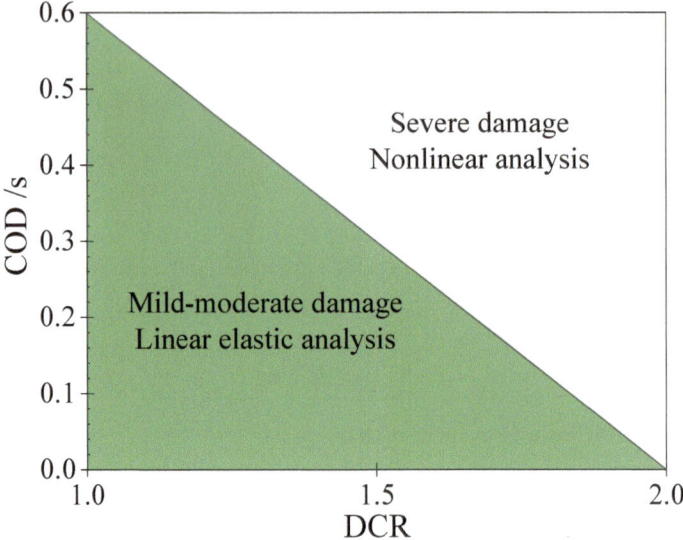

Fig. 8.6 Seismic performance index and failure grade standard of the faced-slabs

mild destruction of the panel is 97.8% and the probability of moderate destruction is 87.1%; when PGA = 0.6 g, the probability of mild destruction of the panel is 100% and the probability of moderate destruction is 94.0%; when PGA = 0.7 g, the probability of mild destruction of the panel is 100% and the probability of moderate destruction is 94.0%. However, the damage of the panel based on the two-dimensional elastic–plastic analysis is not very accurate, and we can only understand its general distribution law and analyze its damage state qualitatively and quantitatively from the perspective of probability.

8.2 Seismic Safety Evaluation of CFRD Considering the Coupled Randomness of Ground Motion and Material Parameters

Table 8.2 presents the performance relationship between various seismic intensity levels, relative dam crest settlement rates, and exceedance probabilities. Figure 8.9 depicts the fragility curve, which serves as a reference for the performance-based seismic safety assessment of CFRDs. It is evident that the difference in failure probabilities between seismic randomness and coupling randomness is within 5%. Hence, seismic randomness primarily governs post-earthquake deformations. In the establishment of a framework for performance-based seismic safety assessment, the randomness of material parameters can to some extent be disregarded.

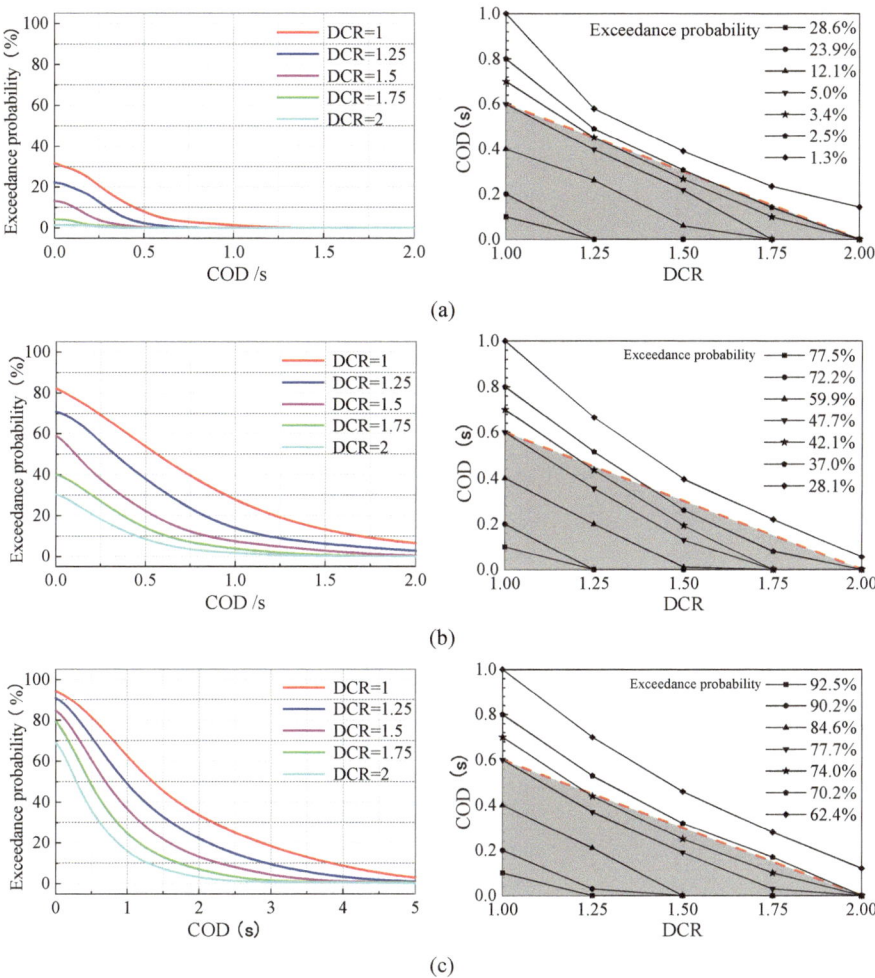

Fig. 8.7 Exceedance probability of faced-slab damage index under different PGA **a** PGA = 0.2 g
b PGA = 0.3 g **c** PGA = 0.4 g **d** PGA = 0.5 g **e** PGA = 0.6 g **f** PGA = 0.7 g

8.3   Seismic Safety Evaluation of 3-D CFRDs Based on Performance

8.3.1   Dam Crest Subsidence

Table 8.3 lists the relationship between multiple earthquake intensities-relative settlement of dam crest-exceedance probability based on 3-D stochastic dynamic response
analysis. Figure 8.10 shows the fragility curves for the performance-based seismic
safety evaluation of CFRDs. There are some differences between the subsidence

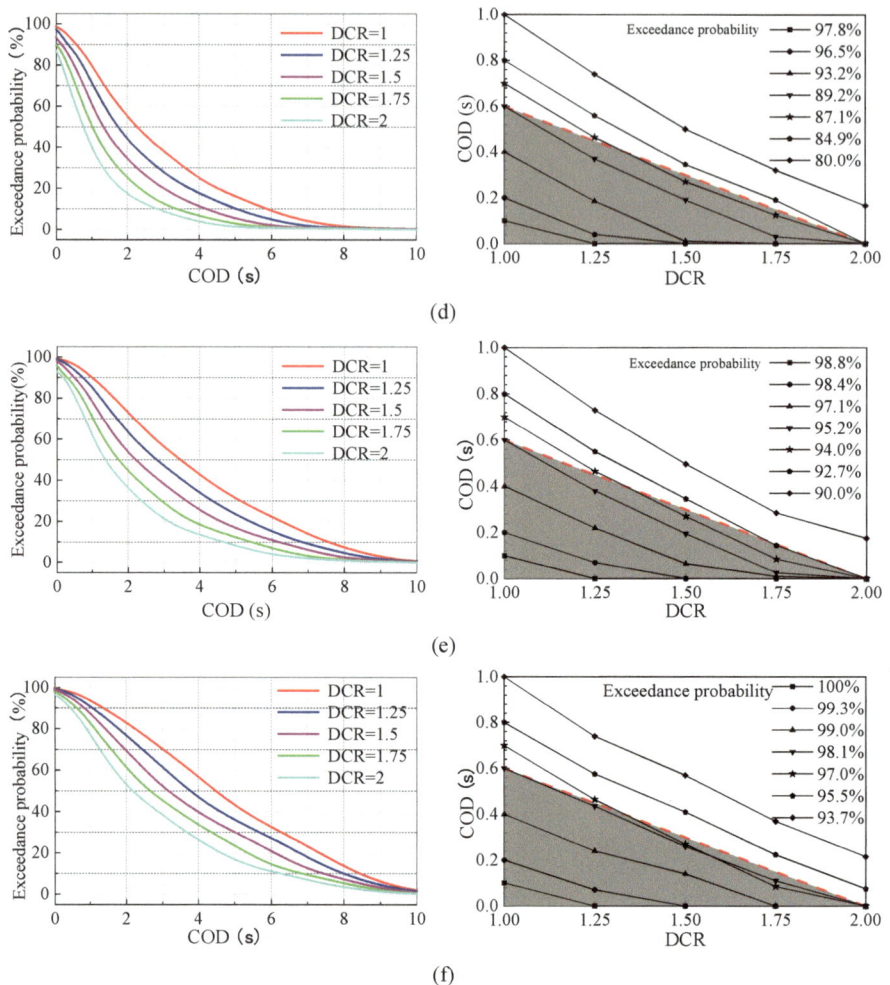

Fig. 8.7 (continued)

values calculated in 2-D and 3-D. The damage probability in 3-D can be obtained by complementing the damage probability in 2-D for different seismic intensities.

8.3.2 Safety of Concrete Slab Impermeable Body

In this paper, the overstress volume ratio of more than 55% or the accumulation time of more than 8 s is taken as the critical state, and faced-slab seismic safety evaluation index and damage grade standard based on the overstress volume ratio and the

Fig. 8.8 Fragility curve based on faced-slab failure

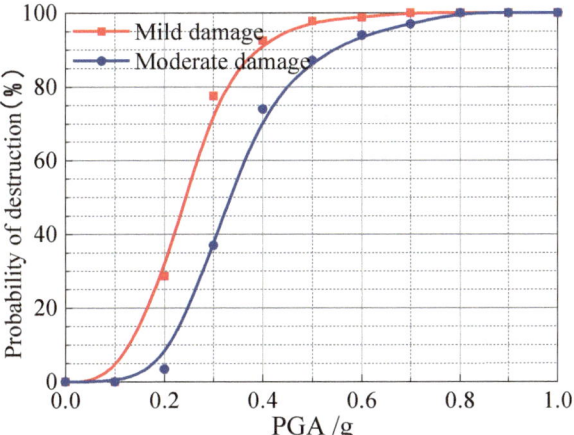

Table 8.2 Relationship of multiple earthquake intensities-relative settlement of dam crest-exceedance probability

Exceedance probability /%	Relative settlement rate of dam crest/%											
	0.1	0.2	0.3	0.4	0.5	0.6	0.7	0.8	0.9	1.0	1.1	1.2
PGA 0.1 g	42.5	0	0	0	0	0	0	0	0	0	0	0
0.2 g	86.3	41.7	8.9	0	0	0	0	0	0	0	0	0
0.3 g	97.2	76.5	42.2	15.2	3.4	0	0	0	0	0	0	0
0.4 g	98.7	90.7	68.6	40.3	17.7	6.0	0	0	0	0	0	0
0.5 g	100	97.4	85.5	63.4	38.3	18.5	6.9	1.8	0	0	0	0
0.6 g	100	100	93.1	78.5	57.5	31.5	18.5	7.6	2.5	0	0	0
0.7 g	100	100	97.1	88.1	71.5	51.0	30.6	15.8	6.5	2.1	0	0
0.8 g	100	100	100	90.8	78.9	62.5	43.1	26.1	13.5	5.7	2.1	0
0.9 g	100	100	100	95.4	86.8	72.5	54.2	35.4	20.1	9.5	4.0	1.3
1.0 g	100	100	100	97.2	91.1	79.5	62.5	43.6	26.5	13.5	6.2	2.4

accumulation time of overstress is established, as shown in Fig. 8.11. Furthermore, Table 8.4 lists the relationship between multiple earthquake intensities-cumulative time-exceedance probability. Figure 8.12 shows state lines under different critical ground motion intensities. The fragility curves of different critical state based on faced-slab damage is shown in Fig. 8.13, while the fragility curve based on relative settlement rate of dam crest is added to refine the evaluation framework as the dotted line. Combining the above research and engineering practice, this paper suggests that the seismic intensity be divided into [0,0.2 g], [0.2–0.5 g], [0.5–0.8 g], [0.8–1.0 g], corresponding to four states of the CFRD being in the "basically intact", "mildly-moderately" damaged, " moderately-severely" damaged, and "rarely failed" states,

Fig. 8.9 Fragility curves based on relative settlement rate of dam crest

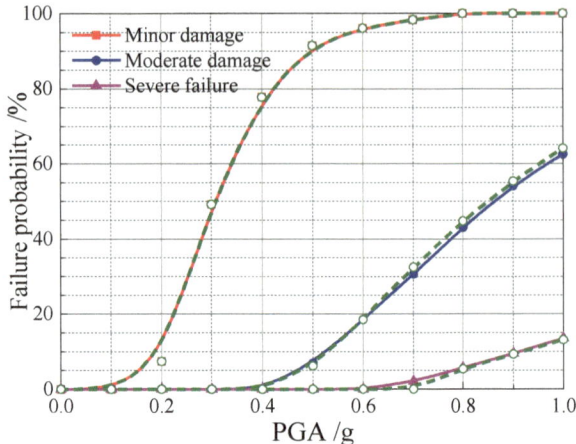

Table 8.3 Relationship of multiple earthquake intensities-relative settlement of dam crest-exceedance probability

Exceedance probability (%)	Crest subsidence ratio (%)									
	0.1	0.2	0.3	0.4	0.5	0.6	0.7	0.8	0.9	1.0
PGA	0.1 g	84.8	3.3	0	0	0	0	0	0	0	0
	0.2 g	100	77.2	15.1	0	0	0	0	0	0	0
	0.3 g	100	98.0	67.1	18.5	1.9	0	0	0	0	0
	0.4 g	100	100	92.3	53.0	14.5	2.0	0	0	0	0
	0.5 g	100	100	100	80.0	38.1	10.2	1.8	0	0	0
	0.6 g	100	100	100	92.0	61.5	25.8	7.5	1.6	0	0
	0.7 g	100	100	100	97.3	78.4	42.4	16.0	5.1	0	0
	0.8 g	100	100	100	100	87.2	57.8	26.9	10.7	3.7	0
	0.9 g	100	100	100	100	92.5	69.2	37.5	16.3	7.1	2.3
	1.0 g	100	100	100	100	93.7	75.4	47.9	24.7	11.5	4.9

respectively. It can be seen that the slab damage of Zipingpu CFRD is in mild-moderate condition, which is consistent with the actual situation. Under mild earthquakes (0–0.2 g), the dam is basically intact, with only a very small number of dams (less than 5 percent) experiencing "mild damage", and basically no damage occurring above the threshold. At the limit of mild damage, the probability of exceeding the PGA of 0.2 g is about 1.5%. Under moderate earthquakes (0.2–0.5 g), the dam is in the state of "mild-moderate" damage, only a very small number (less than 5%) reaches moderate damage. The exceedance probability of moderate dam damage ranges from 1.3% to 3.8% under PGA = 0.5 g, with only a few cases. For severe earthquakes (0.5–0.8 g), the dam is in a state of "moderate-to-severe" damage, with very few (less than 5%) reaching severe damage, and basically no dam failures. The

Fig. 8.10 Fragility curve based on relative settlement rate of dam crest (Dashed lines are based on 2-D analysis)

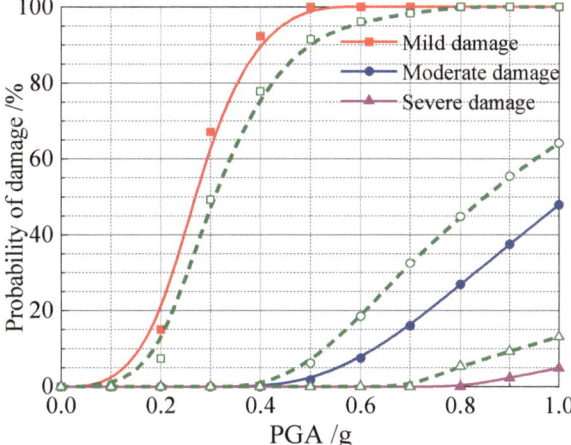

probability of exceeding the threshold of severe damage corresponding to a 0.8 g PGA is 3.7%-6.6%. For larger earthquakes (PGA = 1.0 g), very few (less than 5%) dam failures are allowed, with an exceedance probability of 3.7% for 1.0 g. Earthquakes above 1.0 g are generally less frequently encountered. The result shows that the dam deformation is more likely to reach a mild damage compared to faced-slab damage. Moderate and severe damage is then more likely to be caused by faced-slab damage. The above analyses basically show the reasonableness of the performance standard division criterion in this paper from the perspective of probability. It shows a certain correspondence with the deformation-based division criterion.

Fig. 8.11 Faced-slab seismic safety evaluation index and damage grade standard

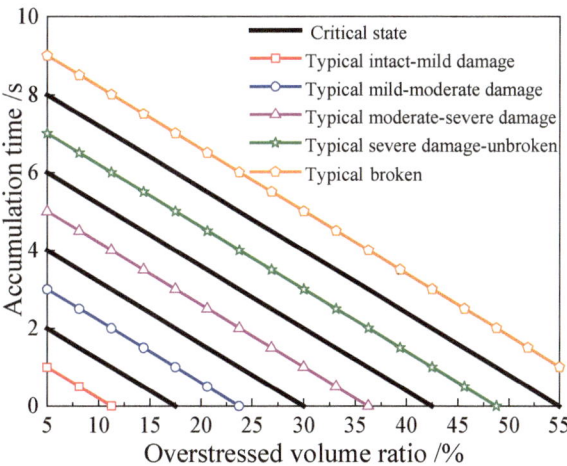

Table 8.4 Relationship table of multiple earthquake intensities-cumulative time-exceedance probability

Exceedance probability (%)		PGA(g)									
		0.1	0.2	0.3	0.4	0.5	0.6	0.7	0.8	0.9	1.0
Accumulation time of overstress (s)	0.5	Intact	11.1	65.9	91.2	97.8	100	100	100	100	100
	1.0		1.5	30.8	75.4	91.2	96.9	100	100	100	100
	1.5		0	11.2	50.2	80.0	92.3	98.1	100	100	100
	2.0			1.9	26.0	60.8	83.1	93.9	98.1	100	100
	2.5			0	10.5	39.1	67.7	86.2	94.3	98.4	100
	3.0				3.3	21.7	48.1	73.6	87.1	95.2	98.4
	3.5				0	10.3	30.3	56.9	76.0	88.7	95.3
	4.0					3.8	17.5	39.4	61.8	78.5	89.5
	4.5					1.3	8.8	24.8	46.7	65.4	80.4
	5.0					0	3.7	14.3	32.7	51.4	68.4
	5.5						1.4	7.6	20.9	38.3	55.1
	6.0						0	3.8	12.1	26.8	42.4
	6.5							2.0	6.6	17.5	31.3
	7.0							0	3.7	10.6	22.0
	7.5								2.1	6.2	14.8
	8.0								1.4	3.6	9.5
	8.5								0	2.2	5.9
	9.0									1.6	3.7

8.4　Performance-Based Safety Evaluation for Dam Slope Stability

8.4.1　Cumulative Time of $F_S < 1.0$

In seismic design of earth-rock dams in China, the pseudo-static method occupies an important position and is widely used with consensus on judgment criteria. When the safety factor Fs is less than 1.0, slope instability is indicated. However, this method is no longer suitable for high-intensity areas with high earth-rock dams. Currently, the finite element dynamic time-history method is commonly used, but there are significant differences in evaluation criteria. Li et al. (2010) suggested that when the cumulative time of Fs < 1.0 exceeds 2 s, slope instability occurs. Zhao et al. (2015) argue that using the dynamic time-history method, when Fs < 1.0, slope instability is observed. When using the dynamic equivalent value method, Fs < 1.1 indicates slope instability. Chen et al. (2013) proposed that when the cumulative time of Fs < 1.0 exceeds 2 s, slope instability occurs. Tian et al. (2014) recommend that when conducting numerical analysis of slope stability, the maximum thickness of

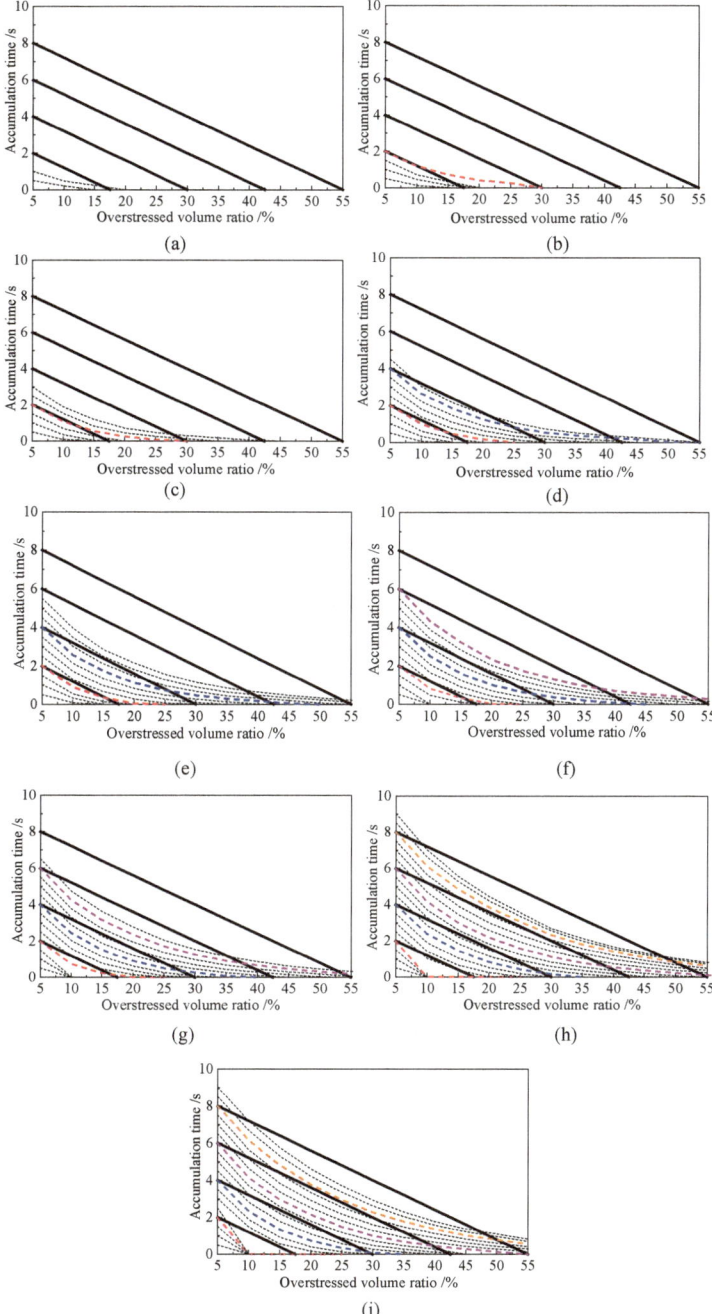

Fig. 8.12 State lines under different ground motion intensities **a** PGA = 0.2 g **b** PGA = 0.3 g **c** PGA = 0.4 g **d** PGA = 0.5 g **e** PGA = 0.6 g **f** PGA = 0.7 g **g** PGA = 0.8 g **f** PGA = 0.9 g **h** PGA = 1.0 g

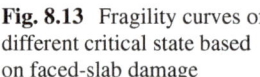

Fig. 8.13 Fragility curves of different critical state based on faced-slab damage

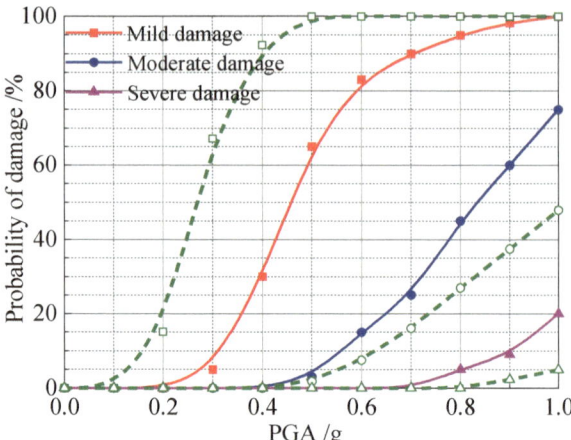

the sliding block should not be less than 5 m, and when the cumulative time of Fs < 1.0 exceeds 1 s, slope instability occurs. Some domestic and international studies and standards also use cumulative slip displacement as an indicator to evaluate slope stability. Ozkan (1998) set a seismic slip deformation limit of 1 m. Switzerland adopts a two-level seismic design criterion (Darbre 2004), allowing deformations of 20 cm for shallow sliding and 50 cm for deep sliding. Tian et al. (2013) propose that when the cumulative slip displacement exceeds 1 m or exceeds 1% of the length of the sliding body, slope instability may occur. Shao et al. (2011) indicate that when the slip displacement exceeds 2% of the length of the sliding body, slope instability occurs.

Figure 8.14 shows the relationship between the average cumulative time between PGA. As the seismic intensity increases to a certain threshold, the dam slope sliding suddenly occurs and there is a significant increase in cumulative time at a specific PGA, followed by a stable increase. This indicates that when the seismic intensity reaches a certain level, the dam slope becomes completely instability. Based on the relationship between the average cumulative time and PGA, the performance boundaries were determined combined with the discussion in Sect. 1.5 using a turning point method. The preliminary classification criteria for failure levels are as follows: when the safety factor is less than 1.0, indicating the onset of cumulative time which is corresponding to "sliding critical" in Table 8.5; mild to moderate damage (0–0.4 s), with 0.5 s as the boundary for moderate damage; cumulative time of 1.5 s as the boundary for severe damage; and 2 s is suggested to be the boundary of partial dam break. Therefore, it is preliminarily recommended to use the cumulative time thresholds of 0 s, 0.5 s, and 1.5 s as performance levels for the classification of high CRFD slopes damage, corresponding to mild, moderate, and severe damage levels, respectively. Furthermore, a multiple earthquake intensities-cumulative time-exceedance probability relationship table was established in Table 8.5, and the corresponding fragility curve is shown in Fig. 8.15, providing valuable references for performance-based dam slope stability and safety design.

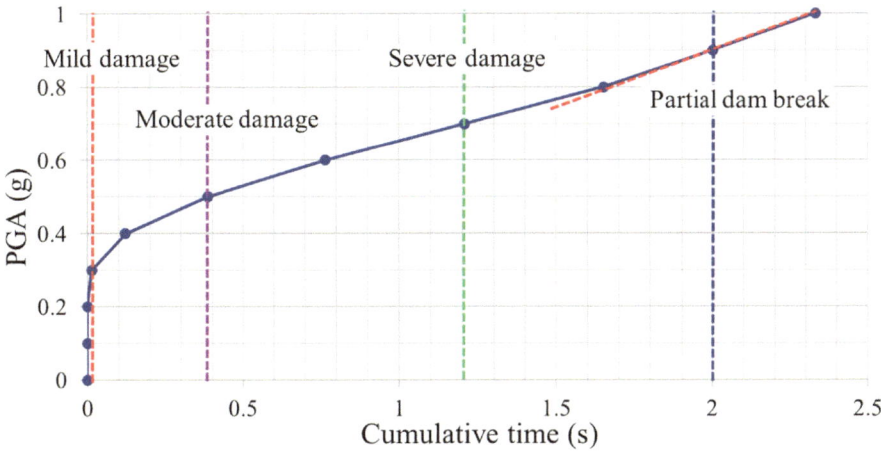

Fig. 8.14 Relationship curve between cumulative time and PGA

Table 8.5 Relationship of multiple earthquake intensities-cumulative time-exceedance probability

Exceedance probability (%)		Cumulative time (s)									
		0	0.05	0.1	0.5	1	1.2	1.5	2	2.5	3
PGA	0.1 g	0	0	0	0	0	0	0	0	0	0
	0.2 g	0	0	0	0	0	0	0	0	0	0
	0.3 g	10.7	8.4	5.7	0	0	0	0	0	0	0
	0.4 g	50.7	45.1	36.7	4.6	0	0	0	0	0	0
	0.5 g	83.1	78.6	71.3	26.7	10.5	6.2	2.5	0	0	0
	0.6 g	95.7	93.9	90.9	58.8	31.5	22.8	12.4	3.1	0	0
	0.7 g	100	100	98.2	82.1	62.4	51.2	32.5	11.5	2.5	0
	0.8 g	100	100	100	95.2	82.8	74.5	59.3	30.4	9.6	1.7
	0.9 g	100	100	100	98.3	91.2	87.0	77.2	51.5	23.0	7.0
	1.0 g	100	100	100	100	97.2	94.8	88.7	68.0	39.1	16.5

8.4.2 Cumulative Slippage

Figure 8.16 shows the relationship between the cumulative slippage and PGA. As the seismic intensity increases to a certain threshold, the dam slope sliding suddenly occurs, and there is a significant increase in cumulative slippage at a specific PGA, followed by a stable increase. This indicates that when the seismic intensity reaches a certain level, the dam slope becomes completely instability. Based on the relationship between the average cumulative slippage and PGA, the performance boundaries were determined combined with the discussion in Sect. 1.5 using a turning point method. The preliminary classification standards for damage levels are as follows: mild damage occurs when the safety factor is less than 1.0, corresponding to "minimal

Fig. 8.15 Fragility curve based on cumulative time

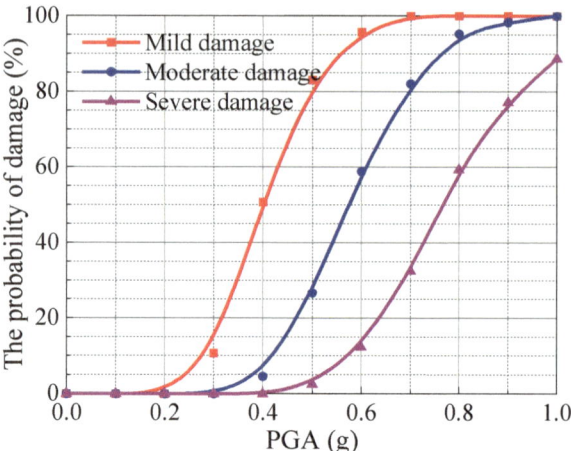

sliding" in Table 8.6; mild to moderate damage (0–15 cm), with 20 cm as the moderate damage limit; 100 cm as the severe damage limit, which is also the classification criteria commonly used in the analysis of Sect. 1.5; and it is suggested to set 150 cm as the criterion for partial dam break. Therefore, it is preliminarily recommended to use cumulative slippage of 0, 20, and 100 cm as performance thresholds for high CRFD failure, corresponding to mild, moderate, and severe damage levels, respectively. It can be observed that this classification method is somewhat similar to that established based on the cumulative time. A multiple earthquake intensities-cumulative slippage-exceedance probability relationship table was established in Table 8.6, providing valuable references for performance-based dam slope stability and safety design. Additionally, it can be seen that both of the two methods based on the cumulative time and the cumulative slippage exhibit a certain degree of correlation from Fig. 8.17.

8.5 Dam Slope Stability Performance Evaluation Considering Coupling Randomness of Ground Motion-Shear Strength Parameters

8.5.1 Basic Information

In this section, the coupling randomness of seismic motion and shear strength parameters are considered and 144 sets of non-stationary acceleration time histories as well as the random shear strength parameters φ_0 and $\Delta\varphi$ are generated. The working conditions and load conditions of the high CFRD are the same as those in Sect. 3.4.1. The static and dynamic parameters are adopted those in Sect. 7.5.1. 1440 acceleration time

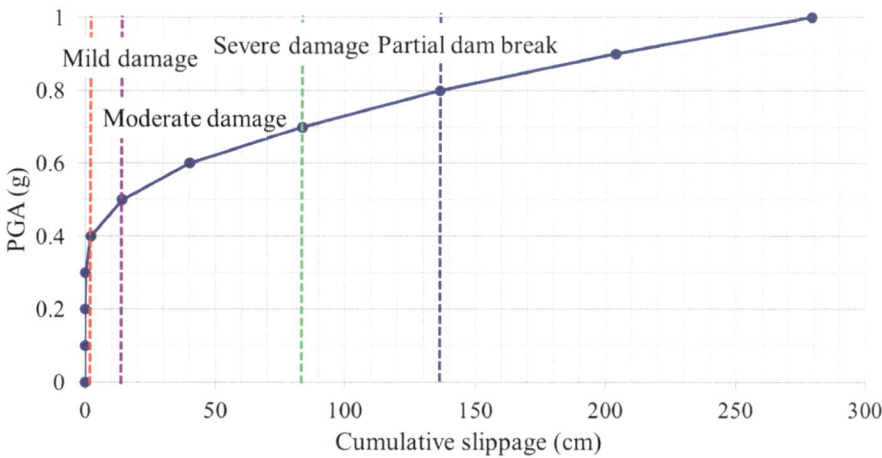

Fig. 8.16 Relationship curve between cumulative slippage and PGA

Table 8.6 Relationship of multiple earthquake intensities-cumulative slippage-exceedance probability

Exceedance probability (%)		Cumulative slippage (cm)										
		0	1	5	10	20	50	80	100	150	200	250	300
PGA	0.1 g	0	0	0	0	0	0	0	0	0	0	0	0
	0.2 g	0	0	0	0	0	0	0	0	0	0	0	0
	0.3 g	10.0	2.5	0	0	0	0	0	0	0	0	0	0
	0.4 g	40.0	24.6	9.6	5.4	2.8	0	0	0	0	0	0	0
	0.5 g	67.4	55.6	38.0	28.8	20.8	8.3	3.7	2.5	0	0	0	0
	0.6 g	84.3	79.2	64.9	54.3	44.3	29.9	18.2	12.6	4.8	2.7	0	0
	0.7 g	94.1	91.9	85.2	79.2	70.9	52.5	40.0	33.8	19.6	9.2	5.6	3.2
	0.8 g	97.6	96.1	94.8	92.1	87.1	72.3	60.4	53.7	38.1	26.1	15.6	8.7
	0.9 g	100	100	97.7	96.5	93.9	85.9	76.2	69.5	55.2	43.9	33.3	23.6
	1.0 g	100	100	100	100	98.4	94.4	88.9	84.4	71.2	58.6	48.5	39.8

histories were generated, with PGA $= 0.1$ to 1.0 g with a 0.1 g interval, and each level had 144 stochastic ground motions. Subsequently, the GPDEM, reliability probability analysis, and fragility analysis are applied to study the impact of coupling randomness on the stability of high CFRD from a perspective of stochastic dynamics and probability. This further improves the performance-based seismic safety evaluation system for dam slope stability.

Fig. 8.17 Fragility curves based on cumulative slippage

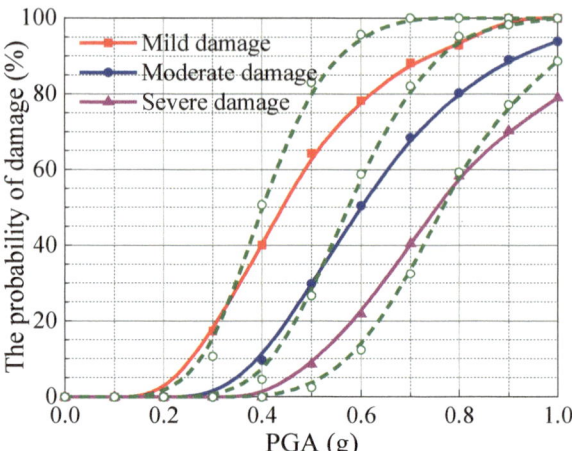

8.5.2 Safety Factor

Figure 8.18 illustrates the time history of mean and standard deviation of the safety factor with PGA = 0.5 g, which shows that the influence of the coupling randomness of seismic motion and material parameters on the safety factor is essentially consistent with the influence of seismic motion randomness. However, it differs significantly from the influence of material parameter randomness. The discrete distribution and exceedance probability of the minimum safety factor with PGA = 0.5 g shows that the response caused by the coupling randomness has minor differences compared to the response induced by seismic motion randomness, but substantial differences compared to the response affected by material parameter randomness, seen from Fig. 8.19. These results demonstrate that seismic motion randomness plays a dominant role in the safety factor response.

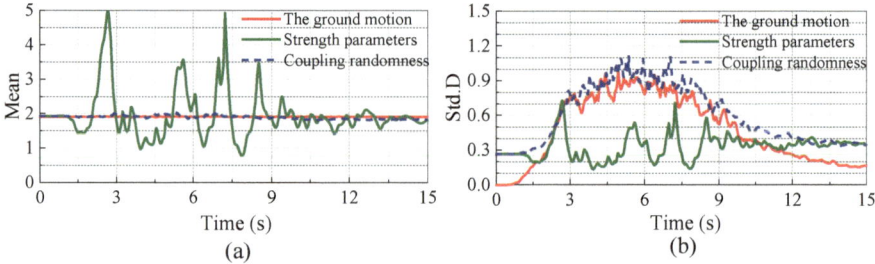

Fig. 8.18 Mean and standard deviation time history of safety factor **a** mean **b** standard deviation

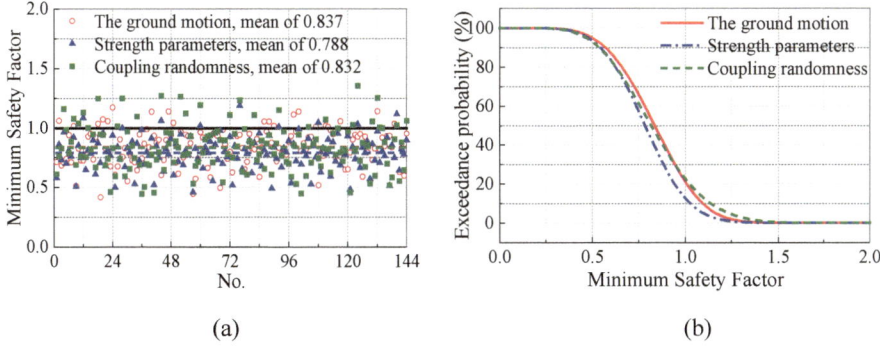

Fig. 8.19 Distribution and exceedance probability of minimum safety factor **a** distribution **b** exceedance probability

8.5.3 Cumulative Time of $F_S < 1.0$

Figure 8.20 illustrates the discrete distribution and exceedance probability of the cumulative time of $F_S < 1.0$ with PGA $= 0.5$ g and it is evident that the coupling randomness has a significant impact on the cumulative time of $F_S < 1.0$, showing notable differences compared to the influence of seismic motion randomness alone. This indicates that the shear strength parameters have a substantial effect on the cumulative time of $F_S < 1.0$. Therefore, it is crucial to fully consider the coupling randomness for the performance-based seismic safety evaluation of dam slopes.

Figure 8.21 illustrates the exceedance probabilities of cumulative time of $F_S < 1.0$ with different seismic intensities, mainly distributed between 95 and 5% exceedance probabilities. Under the influence of the coupling randomness, the variability becomes more obvious. The numerical ranges are as follows: 0–0.34 s (0.3 g), 0–1.40 s (0.4 g), 0–2.18 s (0.5 g), 0–2.68 s (0.6 g), 0.06–3.15 s (0.7 g), 0.23–3.58 s (0.8 g),

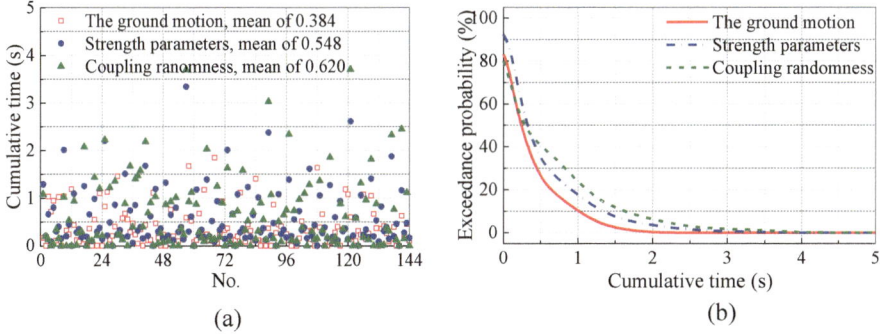

Fig. 8.20 Distribution and exceedance probability of cumulative time **a** distribution **b** exceedance probability

Table 8.7 Relationship table of multiple earthquake intensities-cumulative time-exceedance probability

Exceedance probability (%)		Cumulative time (s)									
		0	0.05	0.1	0.5	1	1.2	1.5	2	2.5	3
PGA	0.1 g	0	0	0	0	0	0	0	0	0	0
	0.2 g	0	0	0	0	0	0	0	0	0	0
	0.3 g	21.7	18.6	14.5	2.8	0	0	0	0	0	0
	0.4 g	51.3	46.8	40.5	17.1	7.0	5.9	4.6	2.5	0	0
	0.5 g	79.4	74.5	67.3	40.5	23.8	18.0	11.9	6.3	3.1	0
	0.6 g	90.8	88.2	83.1	57.5	41.9	36.4	26.7	13.8	6.9	3.0
	0.7 g	96.5	95.4	93.3	75.5	56.9	51.2	42.3	25.4	13.6	6.5
	0.8 g	100	97.3	96.4	89.2	72.1	64.9	55.0	38.8	23.9	12.3
	0.9 g	100	100	100	94.8	85.3	78.7	68.1	51.0	34.5	19.7
	1.0 g	100	100	100	96.3	91.8	87.8	79.9	63.5	45.8	29.1

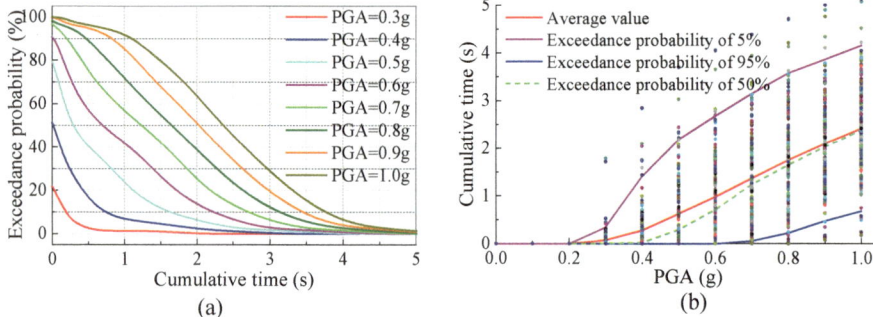

Fig. 8.21 Exceedance probability of cumulative time of $F_S < 1.0$ under different PGA **a** Exceedance probability **b** Cumulative time of $F_S < 1.0$

0.48–3.87 s (0.9 g), and 0.68–4.16 s (1.0 g). These results provide valuable references for the performance-based seismic safety evaluation of dam slopes. Table 8.7 lists the exceedance probabilities for cumulative time under different seismic intensities, and the fragility curve based on the cumulative time of $F_S < 1.0$ is derived, as shown in Fig. 8.22. Notably, there are discernible differences compared to the exceedance probabilities obtained only considering seismic motion randomness.

8.5.4 Cumulative Slippage

Figure 8.23 illustrates the discrete point distribution and exceedance probabilities of cumulative slippage with PGA = 0.5 g, which indicates that the coupling randomness

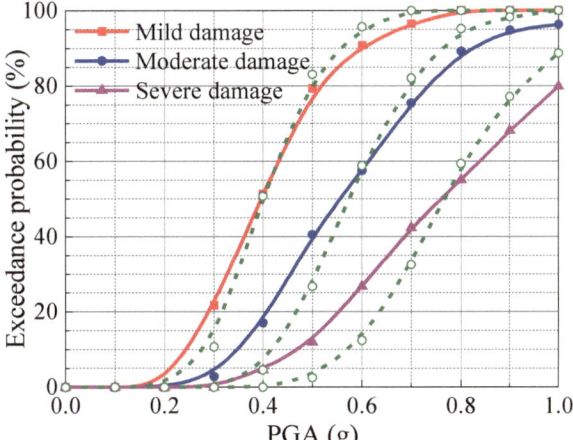

Fig. 8.22 Fragility curve based on cumulative time

of ground motions and material parameters significantly influences the cumulative slippage, showing notable differences compared to the influences of seismic motion randomness and material parameter randomness alone. This indicates that the shear strength parameters have a large effect on the cumulative slippage. Therefore, it is crucial to fully consider the coupling randomness for the performance-based seismic safety evaluation of dam slopes.

Figure 8.24 shows the exceedance probabilities of cumulative slippage and Fig. 8.25 illustrates the cumulative slippage under different seismic intensities, which are mainly distributed between 95 and 5% exceedance probabilities and they exhibit greater variability when considering the coupling randomness. The numerical ranges are as follows: 0–3 cm (0.3 g), 0–45 cm (0.4 g), 0–129 cm (0.5 g), 0–231 cm (0.6 g), 0–335 cm (0.7 g), 0–462 cm (0.8 g), 0–628 cm (0.9 g), and 14–777 cm (1.0 g). These results serve as a reference for performance-based seismic safety evaluations of dam slopes. Table 8.8 lists the exceedance probabilities for cumulative slippage under

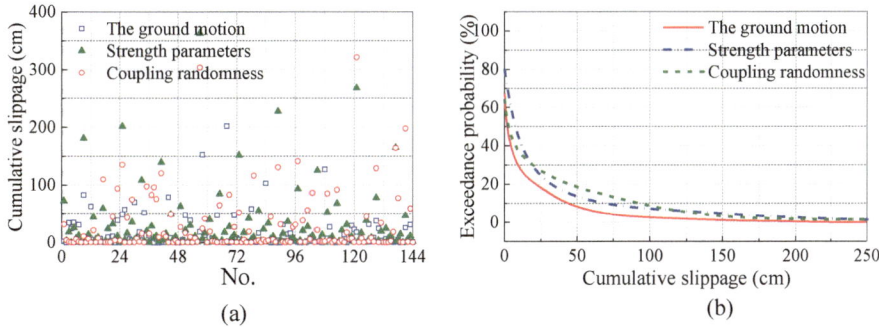

Fig. 8.23 Distribution and exceedance probability of cumulative slippage **a** Distribution **b** Exceedance probability

different seismic intensities, and the fragility curve based on the cumulative time of FS is derived, as shown in Fig. 8.26. Notably, there are discernible differences compared to the exceedance probabilities obtained solely considering the ground motion randomness.

8.5.5 Discussion on the Relationship Between Cumulative Time and Cumulative Slippage

The correlation analysis of cumulative time and cumulative slippage shown in Fig. 8.27 reveals that considering the coupling randomness of ground motion and material parameters results in a stronger correlation between the two factors and a broader range of numerical variability. This indicates the necessity of incorporating the coupling randomness of ground motion and material parameters in the analysis of dam slope stability and seismic safety.

8.6 Conclusion

In this paper, a series of random samples are generated considering the randomness of ground motion, the uncertainty of shear strength parameters and their coupling based on the study of softening effect of rockfill materials. The stochastic dynamic and probabilistic responses of three physical parameters of dam slope stability, safety factor, cumulative time of $F_S < 1.0$ and cumulative slippage under various random factors are analyzed combining with the finite element dynamic time-history analysis method of dam slope stability and GPDEM. The criterion of performance level division based on cumulative time of $F_S < 1.0$ and cumulative slippage is proposed and the performance safety evaluation framework based on multiple earthquake intensities-multiple performance indices-exceedance probability is established. The main work and conclusions are as follows:

(1) First, a step for finite element dynamic stability analysis considering the softening effects of rockfill materials was established. The significant influence of the softening effects on the safety factor, cumulative time of $F_S < 1.0$ and cumulative slippage of the dam slope subjected to earthquakes especially strong ones was revealed through stochastic dynamic and probabilistic analyses. As seismic intensity increases and seismic duration extends, the softening effects become more obvious, exhibiting a progressive process. In addition, the reliability analysis shows that it is unreasonable to evaluate the stability of earth-rock dam slopes only based on the minimum safety factor, and it is necessary to comprehensively assess the seismic safety of dam slopes by combining the cumulative time of $F_S < 1.0$ and the cumulative slippage, which is of great significance for the performance-based evaluation of the seismic safety of dam slopes.

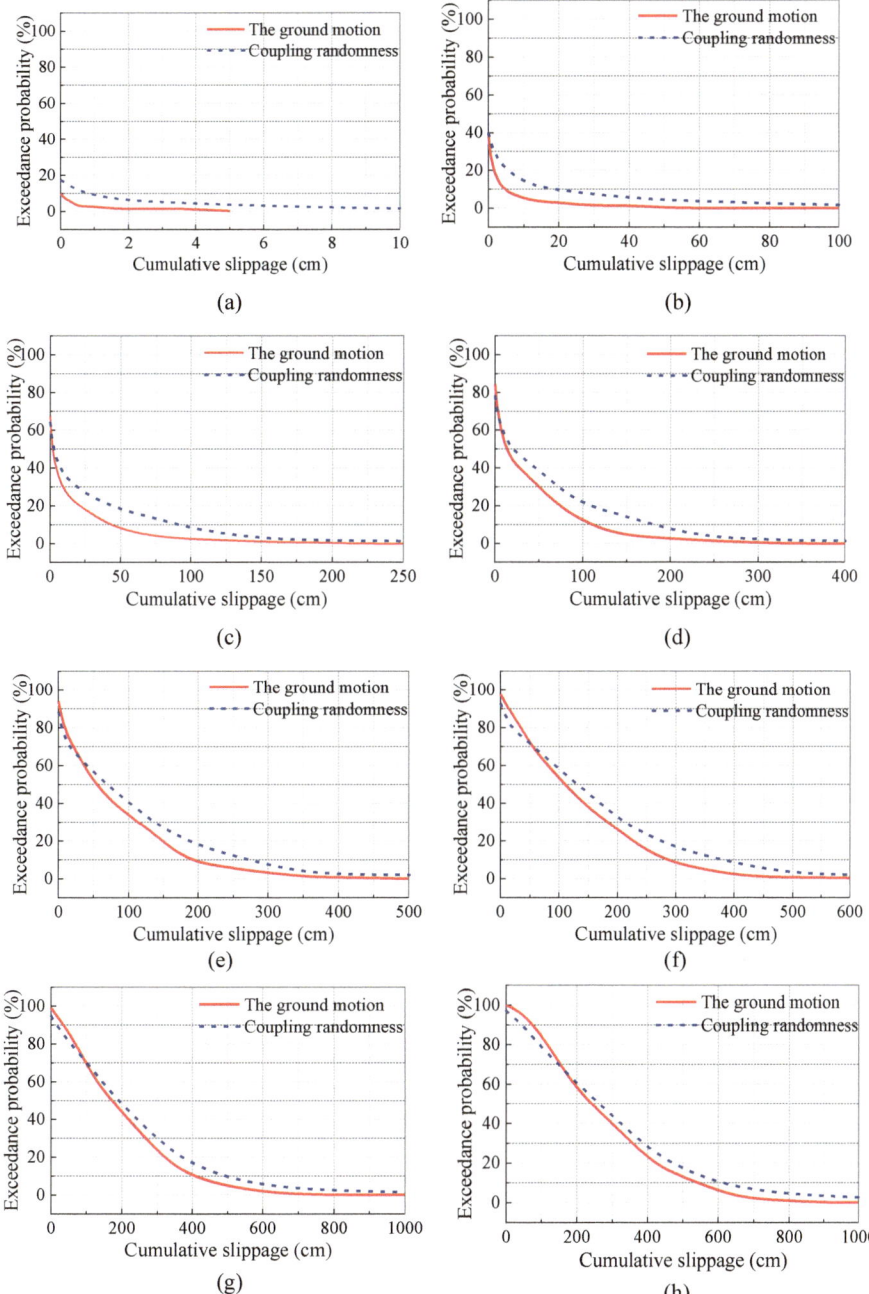

Fig. 8.24 Exceedance probability of cumulative slippage under different PGA **a** PGA = 0.2 g **b** PGA = 0.3 g **c** PGA = 0.4 g **d** PGA = 0.5 g **e** PGA = 0.6 g **f** PGA = 0.7 g **g** PGA = 0.8 g **f** PGA = 0.9 g **h** PGA = 1.0 g

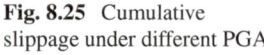

Fig. 8.25 Cumulative slippage under different PGA

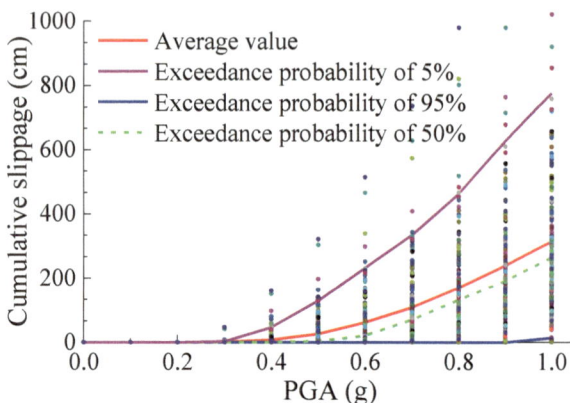

Table 8.8 Relationship of multiple earthquake intensities-cumulative slippage-exceedance probability

Exceedance probability (%)		Cumulative slippage (cm)									
		0	1	5	10	20	50	80	100	150	200
PGA	0.1 g	0	0	0	0	0	0	0	0	0	0
	0.2 g	0	0	0	0	0	0	0	0	0	0
	0.3 g	17.4	9.2	3.7	0	0	0	0	0	0	0
	0.4 g	40.1	33.2	21.0	14.6	9.7	4.5	0	0	0	0
	0.5 g	64.3	56.1	43.5	37.0	29.8	18.5	12.3	8.6	3.3	0
	0.6 g	78.3	75.2	64.1	58.2	50.5	38.5	27.5	21.9	14.2	7.9
	0.7 g	88.2	87.2	79.5	74.2	68.5	56.5	46.8	40.5	27.2	18.2
	0.8 g	92.9	92.2	89.5	83.5	80.3	71.5	63.6	58.3	45.2	33.0
	0.9 g	100	100	93.5	91.5	89.1	81.5	74.9	70.3	58.9	48.3
	1.0 g	100	100	100	96.5	93.9	89.1	82.9	79.0	69.5	60.8

(2) The stochastic dynamic process and probabilistic information of the safety factor, cumulative time of $F_S < 1.0$ and cumulative slippage were obtained from the perspective of ground motion randomness. And it reveals the necessity of analyzing the dam slope stability from a stochastic dynamic viewpoint and indicates a certain correlation between the cumulative time of $F_S < 1.0$ and cumulative slippage. The performance level criteria of cumulative time of $F_S < 1.0$ and cumulative slippage are proposed based on inflection points and relevant researches as follows: cumulative time of 0 s (critical sliding), 0.5 s, and 1.5 s; cumulative slippage of 0 cm (very slight sliding), 20 cm, and 100 cm, corresponding to the boundary states of mild damage, moderate damage, and severe damage, respectively. A relationship table of multiple earthquake intensities-multiple performance indices-exceedance probability was

Fig. 8.26 Fragility curve based on cumulative slippage

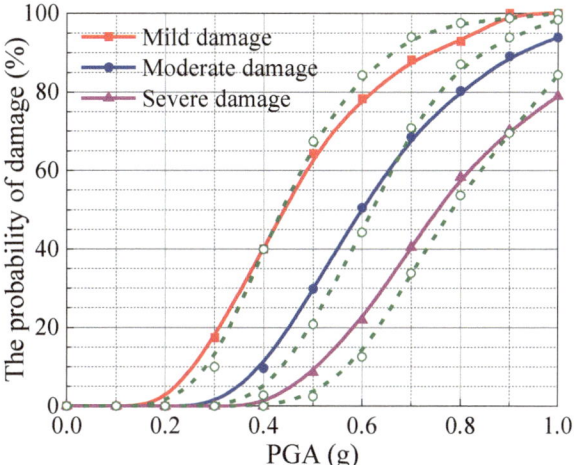

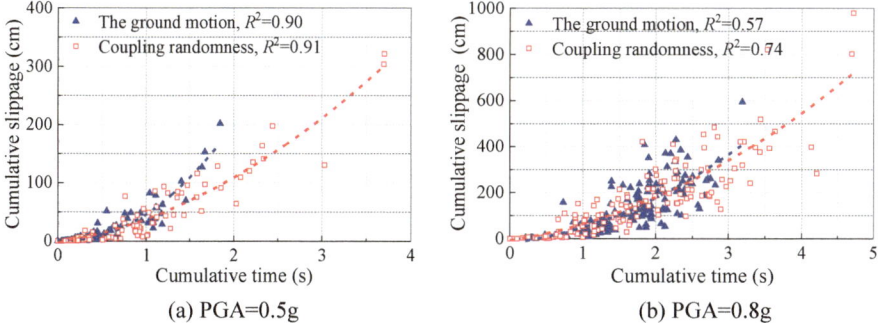

Fig. 8.27 The relationship between cumulative time and cumulative slippage under 0.5 g and 0.8 g **a** PGA = 0.5 g **b** PGA = 0.8 g

established, providing valuable references for performance-based seismic safety assessment of dam slopes and ultimate aseismic capacity analysis.

(3) The stochastic analysis and probability assessment of safety factor, cumulative time of $F_S < 1.0$ and cumulative slippage with PGA = 0.5 g emphasizing the necessity of considering material parameters randomness.

(4) A systematic comparison of the effects of three random cases (random material parameters, random ground motions and their coupling) on the safety factor, cumulative time of $F_S < 1.0$ and cumulative slippage of high CFRD slope stability with PGA = 0.5 g was conducted. The results reveal certain differences in their responses, with stochastic seismic excitation and their coupling showing relatively minor distinctions in the impact on safety factors. A relationship table of multiple earthquake intensities-multiple performance indices-exceedance probability was established and the fragility curves were obtained

based on the proposed damage level criteria considering the coupling random-ness of the material parameter randomness and stochastic seismic excitation, further enhancing the performance-based seismic safety assessment framework for high CFRD.

References

Chen SS, Li GY, Fu ZZ (2013) Safety criteria and limit resistance capacity of high earth-rock dams subjected to earthquakes. Chinese J Geotech Eng 35(01):59–65

Darbre GR (2004) Swiss guidelines for the earthquake safety of dams. Proceedings of 13th WCEE. Vancouver B. C.

Ghanaat Y (2004) Failure modes approach to safety evaluation of dams. In: Proceedings of the 13th world conference on earthquake engineering, p 16

Li GY, Shen T, Zhao KZ (2010) Seismic dynamic behavior and limit seismic analysis on high earth core rock fill dams. Hydro-Sci Eng 01:1–8

Liu J, Liu B, Kong XJ (2012) Estimation of earthquake-induced crest settlements of earth and rock-fill dams. J Hydroelect Eng 03:38–42

Ozkan MY (1998) A review of considerations on seismic safety of embankments and earth and rock-fill dams. Soil Dyn Earthq Eng 17(7–8):439–458

Raphael JM (1984) Tensile strength of concrete. J Proc ACI J Proc 81(17):158–165

Shao L, Chi SC, Li HJ et al (2011) Preliminary studies of ultimate aseismic capacity of high core rockfill dam. Rock Soil Mech 32(12):3827–3832

Shen HZ, Zhang CH, Kou LH (2007) Function-based seismic damage evaluation model of concrete gravity dams. J Tsinghua Univ Nat Sci Edn 47(12):2114–2118

Swaisgood JR (2003) Embankment dam deformations caused by earthquakes. Pacific Conference on Earthquake Engineering

Tian JY, Liu HL, Wu XY (2013) Evaluation perspectives and criteria of maximum aseismic capability for high earth-rock dam. J Disaster Prevention Mitigation Eng 33(S1):128–131

Zhao JM, Liu XS, Yang YS et al (2015) Criteria for seismic safety evaluation and maximum aseismic capability of high concrete face rockfill dams. Chin J Geotech Eng 37(12):2254–2261

Zhou Y (2012) Seismic damage analysis of Zipingpu panel rockfill dam and panel seismic countermeasures for the Wenchuan earthquake. Dalian University of Technology, Dalian

Chapter 9
Conclusion and Prospects

9.1 Conclusion

In this study, an examination of the impact of uncertain factors on the seismic response of high CFRDs is conducted within the framework of stochastic dynamics. A comprehensive performance-based seismic safety evaluation framework is delineated. The investigation encompasses the stochastic nature of ground motion during seismic events, the inherent uncertainties associated with material parameters, and the interconnected randomness inherent in both ground motion and material parameters. The study proceeds by systematically addressing these uncertainties. A stochastic ground motion model, predicated upon the seismic specification spectrum of hydraulic engineering, is meticulously formulated. Furthermore, methodologies for generating high-dimensional random parameter samples and coupled random samples encompassing both ground motion and material parameters are established. Integration with a refined nonlinear finite element dynamic time history analysis method, generalized probability density evolution method, and vulnerability analysis method facilitates the elucidation of seismic dynamic response characteristics pertaining to high-faced rockfill dams. This exploration is conducted through the lens of stochastic dynamics and probability, considering key facets such as dam deformation, impermeable body safety, and dam slope stability. The introduction of a seismic safety evaluation performance index, coupled with the stipulation of corresponding performance levels and their associated probabilities, forms an integral part of the analytical process. Subsequently, the establishment of a multi-seismic intensity-multi-performance target-failure probability performance relationship contributes to the initial formation of a performance-based seismic safety evaluation framework. In summation, the principal findings of this investigation are outlined as follows:

© The Author(s) 2025
B. Xu and R. Pang, *Stochastic Dynamic Response Analysis and Performance-Based Seismic Safety Evaluation for High Concrete Faced Rockfill Dams*,
Hydroscience and Engineering, https://doi.org/10.1007/978-981-97-7198-1_9

(1) Building upon a summary of commonly employed probability analysis methods, this study introduces the Generalized Probability Density Evolution (GPDE) method into the realm of random dynamics and probability analysis for high-faced rockfill dams. Firstly, it is emphasized that under seismic actions, high-faced rockfill dams exhibit numerous sources of uncertainty, primarily stemming from the stochastic nature of seismic motion and the uncertainty in material parameters. Notably, there is a scarcity of research that analyzes the seismic response of high-faced rockfill dams from the perspectives of random dynamic time history and probability. Subsequently, a critical review of prevalent methodologies such as the first- and second-order moment method, Monte Carlo method, and response surface method is undertaken, highlighting their limitations in probabilistic seismic analysis for complex geotechnical engineering, including high-faced rockfill dams. To address these shortcomings, the study proposes the Generalized Probability Density Evolution method based on stochastic dynamics theory, offering applicability to the probabilistic seismic response and analysis of complex geotechnical engineering structures, including high-faced rockfill dams. Finally, the application and solution processes of the Generalized Probability Density Evolution method are elaborated upon. This includes the establishment of a fully non-stationary random seismic motion model based on spectral representation-random function and a high-dimensional random sample parameter generation method employing the GF-deviation optimization point selection technique. Through analytical solutions, Duffing oscillator, multi-layered rock-soil slopes, and equivalent linear random dynamic and probabilistic analyses of structures such as high-faced rockfill dams, the effectiveness and reliability of the combined GPDE method with stochastic seismic motion and high-dimensional random parameter generation methods are validated. This contributes to laying the foundation for the random dynamic analysis of high-faced rockfill dams and performance-based seismic safety assessment in complex geotechnical engineering.

(2) Taking into full consideration the stochastic nature of seismic motion, this study unveils the random dynamic characteristics of high-faced rockfill dams and establishes a performance-based seismic safety assessment process. Initially, a comprehensive set of seismic acceleration time histories with complete probability information is discretely generated. A series of elastic–plastic finite element dynamic analyses is conducted, and in conjunction with the Generalized Probability Density Evolution theory, the random seismic response and extensive probability information of high-faced rockfill dams are obtained. Subsequently, from the perspectives of random dynamics and probability, the study elucidates the variations and distribution characteristics of dam body acceleration, deformation, and panel stress. It indicates that different seismic intensities and seismic actions have a significant impact on the seismic response of high-faced rockfill dams. For a given set of seismic actions, the maximum response can be three to five times the minimum response. Therefore, a comprehensive assessment of seismic safety for high-faced rockfill dams is essential from the viewpoint of seismic motion randomness. The study also reveals the

spatially irregular flow characteristics of the probability responses and the non-normal distribution features, such as non-Gaussian distributions. Based on the 5%, 50%, and 95% exceedance probabilities, the study obtains the ranges of values for various physical quantities, providing reference values for numerical calculations and the analysis of the ultimate seismic capacity of high-faced rockfill dams. Finally, performance level division standards are proposed based on dam body deformation and panel impermeable body safety. These standards are validated for their reasonability through probability assurance. Specifically, for dam body deformation, the performance indicators of 0.3%, 0.7, and 1.0% relative settlement at the dam crest correspond to slight damage, moderate damage, and severe damage threshold states, aligning well with the damage observed in the Zipingpu face rockfill dam during the Wenchuan earthquake. A standard of 1.1% is suggested as the non-failure criterion. For panel impermeable body safety, a dual-control performance indicator is proposed, combining the demand stress ratio with the cumulative over-stress duration. The study establishes a multi-seismic intensity-multi-performance target-exceedance probability performance relationship, constructs vulnerability curves, and outlines a preliminary performance-based seismic safety assessment method for high-faced rockfill dams.

(3) A high-dimensional elastoplastic random parameter sample generation method is established to reveal the stochastic dynamic response and probabilistic characteristics of high-faced rockfill dams under the influence of random material parameters. Initially, elastoplastic random parameter samples are generated based on the GF-deviation optimization point selection method. In conjunction with the Generalized Probability Density Evolution method, the study, from both the perspectives of random dynamics and probability, asserts that under deterministic seismic actions (PGA $= 0.5$ g), the uncertainty in material parameters significantly impacts the seismic response of high-faced rockfill dams, affecting dam body acceleration, deformation, and panel stress. Subsequently, using the 5% exceedance probability and 95% exceedance probability as benchmarks, it is noted that the maximum response is approximately 1.5–2 times the minimum response, indicating a relatively minor impact on seismic response variability compared to the randomness induced by seismic motion. Lastly, through a comparison of response values considering material parameters with normal distribution and log-normal distribution, the study indicates that the distribution type has a negligible effect on seismic response, with differences generally within 10%. Consequently, under deterministic seismic excitation, considering the stochastic nature of material parameters is essential, but the influence of distribution type needs to be considered judiciously.

(4) The seismic motion-material parameter coupling randomness is systematically considered, and its impact on the dynamic response and seismic safety of high-faced rockfill dams is comprehensively studied from the perspectives of random dynamics and probability. This enhances the performance-based seismic safety evaluation framework. Initially, utilizing spectral representation-random function and random material parameter variables, random seismic acceleration time

histories and random material parameter samples are concurrently generated. Subsequently, a detailed investigation from the viewpoints of random dynamics and probability is conducted on the seismic response of high-faced rockfill dams under the influence of different seismic intensities, considering the coupling randomness of seismic motion and material parameters. Through a comparative analysis of the effects of seismic motion randomness, material parameter randomness, and coupled randomness on dam body acceleration, deformation, and panel stress, it is evident that seismic motion primarily controls seismic response. Therefore, in a comprehensive analysis of the random dynamic response of dams, the influence of material parameter randomness can be to some extent neglected. Finally, a performance relationship is established under coupled random factors, encompassing multiple seismic intensities, multiple performance objectives, and exceedance probabilities. Vulnerability curves for different damage levels are obtained, indicating that the differences in performance safety assessment probabilities obtained from seismic motion randomness and coupled randomness are minimal. In the context of performance-based seismic safety assessment for high-faced rockfill dams, consideration of seismic motion randomness alone is deemed sufficient to meet the required criteria.

(5) From the perspective of three-dimensional elastoplastic analysis, this study investigates the performance indicators and performance levels for panel failure evaluation based on the ratio of accumulated overstress volume combined with the cumulative overstress time. This contributes to the further refinement of the performance-based seismic safety evaluation framework. Initially, building upon the previous research findings and considering the stochastic nature of seismic motion, the study, from the standpoint of random dynamics, elucidates the three-dimensional effects on dam body acceleration, deformation, and panel stress. This underscores the necessity of considering seismic motion randomness in seismic response analysis and highlights the significance of response distribution patterns and ranges for the seismic safety assessment and ultimate seismic capacity analysis of high-faced rockfill dams. Subsequently, the study explores and suggests preliminary performance indicators and performance levels for panel seismic safety evaluation based on the ratio of accumulated overstress volume combined with cumulative overstress time. The rationality of performance level division is demonstrated through probability assurance. Finally, the study reveals the probability relationship between two-dimensional deformation and three-dimensional deformation, obtaining vulnerability curves for each performance indicator. This further enhances the performance-based seismic safety evaluation framework, providing a scientific basis for the seismic design and performance control of high-faced rockfill dams.

(6) This study systematically considers the randomness of seismic motion, the uncertainty of material parameters, the coupling randomness of seismic motion and material parameters, and the combined effect of rockfill softening. It explores the performance-based seismic safety evaluation framework for the seismic stability of high-faced rockfill dam slopes from the perspectives of

random dynamics and probability. Firstly, through stochastic dynamic and probability analyses, the study reveals that seismic events, especially strong earthquakes, significantly impact the safety factor, cumulative time exceeding the safety factor, and cumulative slip of dam slopes due to the rockfill softening effect. As seismic intensity and duration increase, the softening characteristics gradually become more pronounced, presenting a progressive process. Reliability analysis indicates that evaluating the stability of soil-rock dam slopes solely based on the minimum safety factor is unreasonable. Instead, a comprehensive assessment of slope seismic safety is necessary, considering cumulative time exceeding the safety factor and cumulative slip. This approach holds crucial significance for performance-based seismic safety evaluation of high-faced rockfill dams. Subsequently, the study uncovers the influence and relationships of seismic motion randomness, material parameter uncertainty, and seismic motion-material parameter coupling randomness on the safety factor, cumulative time exceeding the safety factor, and cumulative slip of dam slopes. The relationship between cumulative time and cumulative slip is also discussed. The necessity of analyzing dynamic slope stability from the perspective of random dynamics and the importance of performance-based seismic safety evaluation are emphasized. This provides valuable insights for dam slope anti-sliding design and the analysis of ultimate seismic capacity. Finally, performance level division standards for two performance indicators, namely cumulative time exceeding the safety factor and cumulative slip, are proposed as follows: cumulative time of 0 s (sliding threshold), 0.5, and 1.5 s; cumulative slip of 0 cm (minimal sliding), 20, and 100 cm. These correspond to the threshold states of slight damage, moderate damage, and severe damage, respectively. The study establishes a multi-seismic intensity-multi-performance target-exceedance probability performance relationship and vulnerability curves. These serve as references for dam slope stability safety assessment, contributing to the further refinement of the performance-based seismic safety evaluation framework for high-faced rockfill dams.

9.2 Prospects

The seismic design and safety assessment of high-faced rockfill dams, especially those with elevated panels, constitute a highly complex subject. Performance-based seismic design represents the future direction of structural seismic design and is currently a focal point in earthquake engineering research. However, research on the performance-based seismic safety assessment of high dams, particularly those with elevated panels, is still in its early stages. This paper, considering various uncertainties under seismic actions, employs advanced nonlinear finite element dynamic time history analysis methods and suitable probability analysis methods to reveal the seismic response patterns of high-faced rockfill dams from the perspectives of random dynamics and probability. The aim is to enhance the framework of performance-based seismic safety assessment. Due to the complexity and comprehensiveness of the issues involved, the research in this paper is still at a preliminary stage. The author believes that, building upon existing research achievements, further work is primarily needed in the following aspects:

(1) In-depth exploration of seismic motion uncertainties and refined identification of strong earthquakes. Seismic motion is a complex load with significant uncertainties. Regardless of seismic motion frequency, peak values, amplitude variations, duration, and the arrangement sequence of different impulses, it is essential to further investigate the uncertainties in seismic motion characteristics. This includes probabilistic insights into its impact on the dynamic response of high-faced rockfill dams, encompassing seismic hazard analyses. Additionally, the development of simple and rational seismic motion parameter indicators is a key focus. Identifying strong earthquakes, such as impulse and sequential earthquakes, which have a substantial impact on high-faced rockfill dams, and exploring their uncertainties in dynamic effects, is another research direction.

(2) Statistical analysis and refined uncertainty analysis of pile material model parameters. Based on relevant experiments and engineering cases, enhancing the statistical characterization of pile material model parameters and establishing their distribution models, exploring the correlation between parameters, and developing methods considering spatial variability of pile material are essential. This aims to provide a more detailed understanding of the impact of pile material parameter randomness and the coupling randomness of seismic motion-material parameters on the dynamic response of high-faced rockfill dams.

(3) Research on fast and refined numerical analysis methods. Nonlinear dynamic time history analysis involves significant computational complexity. Therefore, ongoing efforts are required to develop efficient three-dimensional dynamic nonlinear time history analysis methods for the coupled system of reservoir water, dam, and foundation. This includes detailed analysis, such as the elastic–plastic damage analysis of panel failure, which is of great significance for achieving performance-based seismic safety assessment.

(4) Improvement of performance-based seismic safety assessment methods. Currently, for dam deformation, impermeable body safety, and dam slope stability, general methods are based on the numerical constitutive framework of elastic–plastic or equivalent linear models. There is a need to establish a unified numerical simulation analysis framework. Furthermore, proposing comprehensive performance evaluation indicators and defining performance levels are crucial steps toward establishing and refining the performance-based seismic safety assessment system for high-faced rockfill dams.